EUREKA!

EUREKA!

Mindblowing Science
Every Day of the Year

NewScientist

First published in Great Britain in 2021 by John Murray (Publishers)
First published in the United States of America in 2022
by Nicholas Brealey Publishing
Imprints of John Murray Press
An Hachette UK company

This paperback edition published in 2022

4

A CIP catalogue record for this title is available from the British Library

UK Paperback ISBN 978-1-529-39413-9
UK eBook ISBN 978-1-529-39412-2
US eBook ISBN 978-1-529-39588-4

Typeset in Celeste by Palimpsest Book Production Ltd, Falkirk, Stirlingshire

Printed and bound in Great Britain by Clays Ltd, Elcograf S.p.A.

John Murray policy is to use papers that are natural,
renewable and recyclable products and made from wood grown in sustainable
forests. The logging and manufacturing processes are expected to conform
to the environmental regulations of the country of origin.

John Murray (Publishers)
Carmelite House
50 Victoria Embankment
London EC4Y 0DZ

Nicholas Brealey Publishing
Hachette Book Group
Market Place Center, 53 State Street
Boston, MA 02109 USA

johnmurraypress.co.uk
nbuspublishing.com
newscientist.com

CONTENTS

INTRODUCTION

THROUGHOUT MY CHILDHOOD, at the end of every school year, my parents would buy me and my brother each a large, glossy coffee table book on some topic or other as a reward for our 'hard work'. If I'm honest, I'd have preferred a football or, better still, a new bicycle, but I hope I was nevertheless grateful and appreciative.

The books ranged from encyclopaedias and atlases to stunningly colourful tomes on astronomy, anatomy, dinosaurs, plants, engineering marvels, archaeology and ancient history. We built up quite a library. These were the references I turned to when I wished to satisfy my curiosity about some subject, for this was decades before the Internet and there was no Google to help me. My parents and teachers also did what they could to fill in the gaps, just as parents and teachers have done throughout history the world over. After all, kids ask questions.

But why should this curiosity about the world be limited to inquisitive children? It is often said that we are all born natural scientists, with enquiring minds like sponges that constantly absorb information. But most of us lose this innate curiosity when we reach adulthood; we become complacent, content even, about how much we don't understand and just, well, get on with our lives. But it doesn't have to be this way (and shouldn't be, if you ask me).

There are many ways of defining what it means to think about the world in a scientific way. People often tend to think that scientists are a special breed: very smart, dedicated to their subject and born with a natural ability to see and understand the world differently, like being able to do complex mathematics or to grasp abstract concepts.

None of this is really true. Of course, there are nuanced, technical definitions of the 'scientific method', such as ensuring the reproducibility of experiments, or eliminating bias when you attempt to solve problems.

Without a doubt, these things are important in the practice of science. But there is one overarching characteristic that defines what it means to think 'scientifically'. It is, simply put, to be curious about the world around us.

That doesn't mean everyone should be taking advanced courses in genetics or reading textbooks on quantum theory (although don't let me stop you if you want to!). Science is so much more than hard facts or lessons in critical thinking. It is about acknowledging the wonder and magic of even the most mundane things that we encounter in our everyday lives and wanting to find rational explanations for them. The questions in this treasure trove of a book are brilliant examples of exactly the sort of inquisitiveness I mean.

For decades, the 'Last Word' column in *New Scientist* magazine has provided a weekly cornucopia of fascinating, obscure and entertaining questions posed by readers and answered by other readers. This is just a small selection of the best. What's particularly wonderful is how so many of the questions are ones we would never have dreamed of asking in the first place, but when posed by someone else we simply *have* to know the answer – and then share it with everyone else. Could all the water in the universe put out the sun? I've never considered that before – but as a physicist, I find the explanation on page 109 of what would happen if you tried highly plausible . . .

I've been a scientist and a science communicator all my adult life, so I expect I have a head start on many readers of this book. I was already familiar with some of the questions – yet that won't stop me borrowing the explanations to use myself . . . But there were also a surprising number (many months' worth) that either I only partially understood before or that were completely new to me. And then there were a not-insignificant number of questions to which I thought I knew the answer, only to find I was, in fact, entirely wrong – I won't be admitting which ones though!

That, however, is the true joy of science – it is a never-ending journey of discovery, and finding out what we don't know is half the fun. So whether you're a science enthusiast or just someone curious about the world, I hope you enjoy this book. It's about finding the scientist in all of us, something that comes by constantly asking questions and seeking answers. 'Eureka' to that!

Professor Jim Al-Khalili FRS
September 2021

JANUARY

Can you improve your attention span?

OUR WANDERING MINDS tend to make most of us feel guilty. Jonathan Schooler of the University of California, Santa Barbara has attempted to assess 'normal' levels of mind-wandering in the lab by, for example, getting people to read extracts of Leo Tolstoy's *War and Peace* and interrupting them to ask their thoughts at random intervals. Such studies reveal we spend anywhere between 15 and 50 per cent of the time with our head in the clouds.

Such a lack of focus might seem terribly inefficient, but probably isn't. It's unproductive in the context of whatever you are doing currently, but it's potentially productive in the context of whatever it is you're thinking about. You might be reading a book and thinking about planning a party, and while it's compromising your ability to read the book, you're making progress on the party. There is good evidence that a wandering mind is an evolved trait that helps us to think about and plan for the future – something that also fosters a uniquely human creativity.

Measuring perhaps the closest thing – the ability to stay focused on one task, known as selective attention – involves looking at attention shifts on the millisecond scale, such as asking people to state the colour of shapes as they pop up on a screen while ignoring distractions that pop up at the same time. Such experiments show a lot of variation in selective attention. It's low in kids, perhaps because the developing brain has yet to master control over areas that process incoming sensory information. It then improves until the age of twenty, when it plateaus until middle age, before diminishing once again.

Staying off the booze can improve your attention span – when you consume alcohol, you mind-wander more and notice it less. Technologies that promote thought control could help, too, as could meditation. People who practice mindfulness mind-wander notably less.

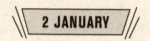

2 JANUARY

Why don't bats get dizzy?

IN HUMANS, MOTION sickness occurs when there is excessive stimulation of the inner ear or from a conflict between sensory information from different sources, such as from the inner ear and the eyes. The parts of the inner ear that are important for orientation with respect to gravity are called the otolith organs: the utricle and the saccule.

The balance mechanism in bats has evolved in a number of ways that allow them to hunt and hang without the problems that humans would face. First, their saccule is slightly rotated forwards, which enables it to act more as a pitch detector, useful in flight. Second, their semicircular canals, which sense rotation of the head, have an internal structure more like a bird's than a human's. This probably allows them to make high-speed turns without the fluid in the canals sloshing back and forth too much. Lastly, if you photograph bats in flight with a high-speed camera, you notice that they keep their heads very stable except in the most violent turns.

But it is the way in which bats sense the world that probably gives them immunity to dizziness. We primarily use vision to do this, but vision is very slow. Anything you look at that takes a second

3

or less to cross 30 degrees of your vision appears smeared. Echolocating bats, while not blind, rely more on biosonar, an especially precise form of hearing that lets them build up 3D images from echoes. Echolocating bats emit brief sonar chirps from 30 to more than 150 times per second, and respond to changes in echoes of less than a microsecond. These bats integrate echolocation with their vestibular system, so they are working with a faster, more precise positioning system than humans do with vision.

Although it is impossible to know for sure whether or not a bat gets dizzy, because it can't tell you if it is, it is possible to infer an animal is dizzy from how it behaves. For example, if an animal is aimlessly walking in circles, fumbling or falling over, it is probably dizzy. Because dizziness and motion sickness usually arise when signals from the vestibular system conflict with those from other sensing systems, bats are less likely to show motion sickness than other mammals.

Am I breathing the same air as Leonardo da Vinci?

THE LAW OF conservation of matter ensures that atoms are constantly being recycled in the Universe. Gravity ensures that most of those on the Earth stay there. So we do indeed breathe in a considerable number of molecules that once passed through Leonardo's lungs – or, unfortunately, Hitler's and anyone else's for that matter.

The total mass of the Earth's atmosphere is about 5×10^{21} grams. If we take air to be a mixture of about four molecules of nitrogen to one of oxygen, the mass of 1 mole of air will be about 28.8 grams. One mole of any substance contains about 6×10^{23} molecules. So there are about 1.04×10^{44} molecules in the Earth's atmosphere.

A single mole of any gas at body temperature and atmospheric pressure has a volume of about 25.4 litres. The volume of air breathed in or out in the average human breath is about 1 litre. So we can assume that Leonardo da Vinci, in one breath, breathed out about 2.4×10^{22} molecules.

The average human takes, say, 25 breaths per minute, so during his 67-year lifetime (1452–1519) Leonardo would have breathed out about 2.1×10^{31} molecules. So, about 1 molecule in every 5×10^{12} molecules in the atmosphere was breathed out by Leonardo da Vinci.

However, because we breathe in about 2.4×10^{22} molecules with each breath, there is a pretty good chance we breathe in about 4.9×10^{9} molecules that Leonardo breathed out. In fact, you can also show in a similar way that you probably breathe in about 5 molecules that he breathed out in his dying gasp.

Of course, there are some pretty crude assumptions involved here in order to arrive at the conclusion. We assume that there has been a good mixing of Leonardo's molecules with the rest of the atmosphere (quite likely in 500 years), that he didn't recycle some of his own molecules, and that there is no loss from the atmosphere due to later users, combustion, nitrogen, fixation and so on. But there is still scope for a considerable loss of molecules without it affecting the main point of the calculation.

By knowing that the number of molecules in the hydrosphere is 5.7×10^{46} molecules, similar calculations can be made for water. These show that a mouthful of liquid contains about 18×10^{6} molecules that passed through Leonardo during his lifetime. So, in addition to breathing in his breath, there is also a pretty good chance of picking up some of Leonardo's urine in every glass of water that you drink.

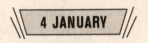

Are rainbows always multicoloured?

A WHITE RAINBOW, known as a fogbow, is formed in essentially the same way as their colourful counterparts.

Sunlight destined to create rainbows and fogbows is refracted twice – once as it enters a water drop and again as it leaves. While inside the water drop between the two refractions, the light bounces off the inside back surface, sending it heading back towards the sun. This is why rainbows and fogbows are seen when the sun is behind the observer. The path of blue light is bent more than red by the droplet, usually causing sunlight to be dispersed into the colours of the visible spectrum with blue at the bottom of a rainbow and red at the top.

Rainbows appear white when water droplets are less than 100 micrometres across – small enough for diffraction to dominate over refraction. Each water drop forms its own diffraction pattern – bands of alternate constructive and destructive interference – for each colour: the smaller the drop, the broader the bands. When the drops are small enough, these bands become so broad that all the colours overlap, essentially mixing them all together again to make white.

Occasionally, fogbows will show a bluish tinge to the inner edge and a reddish one on the outer. From some viewpoints, such as an aircraft, a fogbow can appear as an almost complete circle.

How big is the Internet?

Whatever amount of data the World Wide Web is holding at any one moment, it'll be quite a bit more in the next. A good estimate floating around is 1 yottabyte or a trillion terabytes. Yet this figure dates from 2014, and it's probably safe to assume it's considerably more than that now.

We should also consider the so-called deep web, which encompasses anything that is not found by a mainstream search engine. This includes many large databases for travel bookings, merchandise data for online shopping, any social media networks that do not put everything in the public domain, and so forth. Figures suggest that 80 per cent of the web is dark so if that is true the web comprised 5 yottabytes of data in 2014 and could now hold double that or even more.

It's also important to add that many websites do not reveal how much data they store. Hosting sites hold several redundant copies of almost everything, in multiple locations on varying types of media. Similarly, much of the information on the web is duplicated. Who can guess how many copies of popular content exist?

How do I best surf down an erupting volcano?

You find yourself at the top of an erupting volcano and the only way to save yourself is to surf down a molten lava flow: what should you use to make your surfboard?

Different types of lava melt at different temperatures, rhyolite at up to 900 °C, dacite at up to 1100 °C, andesite at up to 1200 °C and basalt at up to 1250 °C. A lava surfboard must not only be melt-proof but it should also be less dense than the lava underneath and should provide insulation for your feet. Fortunately, volcanoes not only produce lava but also solid materials that are of roughly the same geological composition as lava but are less dense and more insulating because they contain gas bubbles.

A slab of this material, say 50 centimetres thick by 1 metre wide and 2 metres long, would float on molten lava and would melt quite slowly. You could travel a mile or more before you would have to abandon it. Hopefully, by then, you would have been able to negotiate your way to an area of dry, cool ground.

If there happened to be a tree around, you could also use a plank of wood. All woods, and especially oak, form a protective carbonised layer when burnt, which slows further combustion. Timber structures are always designed a little over the size that is actually required so that their structural integrity is retained in case of fire.

Alternatively, if you came prepared and had an old surfboard to hand, you could punch lots of holes through it and connect them to a water tank placed on top of the board. Water escaping through the holes will create the same effect that you can observe when spitting on a hot iron plate: the droplets dance on the plate for quite a long time because they are separated from the plate by a thin layer of steam, which is a bad heat conductor.

This effect would allow you to surf on the lava wave, because the board would be cushioned from the lava by the steam layer, and the friction between the board and the lava would be virtually zero.

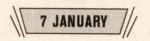

What's the best way to get ketchup out of a glass bottle?

SOME WOULD INSIST that getting ketchup out of a bottle is an art not a science. We disagree for, as you will discover, there seems to be a lot of physics involved. Here we present a few of the best methods supplied to us by *New Scientist*'s readers, each named after the writer:

The Foy (or inertia) method: Most people hit the bottom of the upturned bottle, which only ensures that the inertia of the sauce sends it in the opposite direction – relative to the bottle – to the one you want it to go. The sauce is pushed back into the bottle, rather than out of it. Instead, hold the upturned bottle over your plate and hit the underside of the wrist of the hand holding the bottle with your other fist, jerking the bottle upwards. The inertia of the sauce will now eject it from the bottle.

The Wong (or centrifuge) method: First, put the lid on the bottle and grip it at its base. Then swing your arm as if you were throwing a ball overarm. This method, which uses the principle behind a centrifuge, forces the ketchup to the top of the bottle, allowing you to pour it out. (Whether you can use such a flamboyant technique in a posh chippy is open to question.)

The Lloyd-Evans (or thixotropic) method: Ketchup is gloopy because it is thixotropic. This means that, when it is at rest, it has a thick gelatinous consistency that can be altered to a runny consistency by the input of energy, typically by shaking. The thixotropy is provided by the starch used in ketchup. Starch molecules come in the form of long chains and, when starch powder is mixed with water and heated or subjected to enzyme treatment, weak links are formed between the long molecules. To get the ketchup out of the bottle, first ensure that the lid is on, unless you want to upset the person sitting directly opposite you, then give the bottle some vigorous but not over-athletic shaking. This will break some of those weak bonds between the starch molecules. Now turn the bottle upside down over your plate and watch the ketchup emerge in a slow, gentle stream.

The Medhurst (or 'what-not-to-do') method: Leave the bottle in the back of your store cupboard for a few years until fermentation sets in. Pressure will have built up inside the bottle so that when the lid is removed the sauce will eject dramatically from the opening. Howard Medhurst, who suggested this method, admits that his kitchen now has a 10-cm-wide red streak across the entire ceiling, one wall and half of the floor.

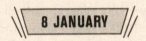

Why do we laugh when we're tickled?

IN 1933, CLARENCE Leuba, a professor of psychology at Antioch College, Ohio planned to find out whether laughter is a learned response to being tickled or an innate one.

To achieve this goal, he determined never to allow his newborn son to associate laughter with tickling. This meant that no one – in particular his wife – was allowed to laugh in the presence of the child while tickling or being tickled. Somehow Leuba got his wife to promise to cooperate, and the Leuba household became a tickle-free zone, except during experimental sessions in which Leuba subjected R. L. Male, as he referred to his son in his research notes, to laughter-free tickling.

During these sessions, Leuba followed a strict procedure. First he donned a 30-centimetre by 40-centimetre cardboard mask, while as a further precaution maintaining a 'smileless, sober expression' behind it. Then he tickled his son in a predetermined pattern – first light, then vigorous – in order of armpits, ribs, chin, neck, knees, then feet.

Everything went well until 23 April 1933, when Leuba recorded that his wife had made a confession. On one occasion, after her son's bath, she had 'bounced him up and down while laughing and saying, "Bouncy, Bouncy".' It is not clear if this was enough to ruin the experiment. What is clear is that by month seven, R. L. Male was happily screaming with laughter when tickled.

Undeterred, Leuba repeated the experiment after his daughter, E. L. Female, was born in February 1936. He obtained the same result. By the age of seven months, his daughter was laughing when tickled.

Leuba concluded that laughter must be an innate response to being tickled. However, one senses a hesitation in his conclusion, as if he felt that it all might have been different if only his wife had followed his rules more carefully. Leuba's tickle study does at least offer an object lesson to other researchers. In any experiment it is all but impossible to control all the variables, especially when one of the variables is your spouse.

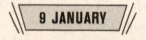

9 JANUARY

What makes the Earth rotate?

EARTH ROTATES SIMPLY because it has not yet stopped moving. The solar system, and indeed the galaxy, were formed by the condensation of a rotating mass of gas. Conservation of angular momentum meant that any bodies formed from the gas would themselves be rotating. As frictional and other forces in space are very small, rotating bodies, including Earth, slow only very gradually.

The moon rotates too, but it presents the same face to us because its period of rotation is the same as its period of revolution around Earth. This equality is the result of tidal friction. If the moon did not rotate, any line through it, parallel to the orbital plane, would keep the same direction in space and the moon would show us its far side during a complete revolution, as one can easily convince oneself by a simple drawing on paper.

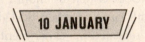

10 JANUARY

Does the sea smell of fish? Or do fish smell of the sea?

THE SMELL OF the seaside is caused by a cocktail of chemicals, principally dimethyl sulphide (DMS), which people can smell at concentrations as low as 0.02 parts per million. Phytoplankton are single-celled organisms found in the sea and use energy from sunlight to make dimethylsulphoniopropionate (DMSP). This chemical is consumed by marine microbes and some is converted into DMS. Because phytoplankton have an earlier origin than fish, and are at the bottom of the food chain, the characteristic seaside smell was around long before fish swam onto the scene. In fact, a newly caught fish rinsed in fresh water does not have a noticeable aroma

DMS molecules act as condensation nuclei for clouds, and James Lovelock was one of those who proposed that it might form part of a negative feedback loop – helping to regulate our climate according to his Gaia hypothesis. As more sunlight reaches Earth, it increases the surface temperature. But it also increases the phytoplankton population so the production of DMS increases and more clouds are seeded. These clouds reflect sunlight, reducing both the surface temperature and the phytoplankton population.

Increased cloudiness is also associated with higher wind speed over the oceans. This mixes surface waters, bringing up nutrients for phytoplankton in readiness for a new burst of sunshine. So when you next stroll along the beach and breathe in the sea air, you might want to reflect on the ocean biochemistry that helps to keep Earth's temperature suitable for life.

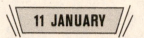

11 JANUARY

Do happy people get ill less?

IT OFTEN SEEMS that people who describe themselves as happy are less likely to catch a cold than those who say they are unhappy. Even when happy people succumb, they seem to have fewer symptoms.

First let's agree that when we talk about 'happiness' we are not referring to transient, hedonistic pleasure, but a general feeling of well-being and satisfaction with life. This varies between individuals and is what relates in some way to a propensity to catch colds and indeed other illnesses. We know there is a correlation, so broadly speaking there are three possibilities: being happy makes you more healthy; being well (more often) contributes to feeling happier; other factors affect mood and overall health.

A study of people after flu vaccinations showed happier folk generated more antibodies. Another study showed that the smiles on photographs of novice nuns were good predictors of their longevity – happier ones living longer. Both of these suggest being happy makes you healthier. We could put forward a mechanism by which happy people socialise more, are exposed to a wider range of pathogens (residing in other people) and so strengthen their immune systems.

There is also a link between infection and mood or depression. This suggests happiness could be the result of not being sick rather than a cause.

The Positive Psychology movement, spearheaded by Martin Seligman, who made his name by studying depression, makes the economic case for paying attention to wellbeing. After all, if increasing happiness means fewer illnesses, less time off work, less pressure on medical resources and so on, that has to be good. And if it's the other way round – no harm done.

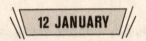

12 JANUARY

Why do all cars smell the same?

'NEW CAR SMELL' is a distinctive smell that seems unchanged over decades and brands. You might be forgiven for thinking that manufacturers secretly spray their cars with the fragrance to seduce new buyers.

The smell actually stems from small molecules called plasticisers, added to the plastics that make up a large proportion of a car's interior. Plasticisers spread throughout the plastic to which they are added, sitting in between the polymer chains so that they can slip more easily over each other. This makes the plastic more flexible and less brittle. However, it is relatively easy for the plasticisers to escape into the air, and in an enclosed environment they can build up to the point that we can easily smell them.

New cars tend to smell the same because they all have roughly the same blend of plasticisers, flame retardants, lubricants and other substances evaporating off interior components and outgassing from dashboard trim, seating foam and upholstery. A more expensive car with leather trim and upholstery is likely to smell a little less of plasticisers and a little more of leather-tanning oils.

The 'new car smell' has actually been synthesised and is available in aerosol cans. It used to be something of a trade secret among car dealers, who would routinely spray the interiors of used cars to make them smell like new. These days, the spray is advertised for sale in car magazines, and can even be bought over the counter at auto parts stores.

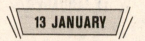

13 JANUARY

How long would it take for a cow to fill the Grand Canyon with milk?

THERE ARE MANY answers to this question, ranging from pedantic arguments over the definition of an average cow to the defeatists' pronouncements that the stench of sour milk would be too great. In the end we decided to treat the answer in its purest form and calculate it based on the volume of the canyon and the average milk output of dairy cattle.

Let's wheel in Daisy, the average cow. In the UK average milk yield per day per cow is in the range 15 to 20 litres. So let's settle on 17.5 litres. The canyon is 446 kilometres long by an average of 16 kilometres wide and 1.6 kilometres deep, which gives a volume of about 10 million billion (10^{16}) litres. So by simple division Daisy would take about 1.8 million million (1.8×10^{12}) years to fill the canyon. This assumes the canyon has a rectangular cross section; for a triangular cross section, the time would be halved.

Now, suppose you don't want to wait 300 times the age of the planet for your canyon full of milk. Instead, you could divert the world's entire milk production to the canyon. This adds another requirement – a milk pumping infrastructure of epic proportions – unless you choose to use dried milk, which would be cheaper to transport, and then rehydrate it with water from the river. The UN Food and Agriculture Organisation estimates that global milk production in 2004 was 504 million tonnes, which is equivalent to 489 billion litres, giving an estimated fill time of only about 20,000 years – still a pretty long job.

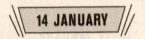

14 JANUARY

Why are tornadoes cone-shaped?

EXACTLY HOW TORNADOES form is not completely understood, but we know that they usually occur during thunderstorms and when volumes of air are unstable. They are the result of updrafts that are created when warm, moist air meets cold, dry air. First, a horizontally spinning column of air called a vortex can form when there are different wind speeds at different altitudes. If this vortex collides with a violent updraft, it can be knocked into an upright position; when the vertically spinning vortex reaches the ground, a tornado is born.

To understand why tornadoes are often shaped like inverted cones, you need to remember that air pressure decreases at higher altitudes. Near ground level, the air surrounding the vortex is at high pressure and crushes the spinning column of air. As the altitude increases, the pressure outside the tornado drops, and it will spread out.

There is more than one way for a tornado to reveal itself, though not all are visible. At the centre of the tornado, the air is spiralling upwards at a great speed and creates a small region of lower pressure. If the pressure there is low enough, water vapour in the air will condense into visible droplets and the tornado will appear as a funnel-shaped cloud.

Sometimes, if the air pressure inside the tornado is not low enough for clouds to form, the tornado will only be revealed by the dirt and debris it picks up from the ground. On the other hand, if the air is very moist and the air pressure extremely low, then the cloud base may be so close to the ground that there is not enough vertical distance for the funnel cloud to taper into a point and the tornado will take the form of a wedge.

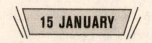

15 JANUARY

Why are goats so smelly?

MALE GOATS ARE notoriously smelly, possessing a peculiar stink that can be quite offensive to the nose.

The malodorous chemical cocktail originates from their urine and from scent glands near their horns. It is so potent that it can bring females into oestrous or sexual receptiveness. In one study, Japanese scientists put hats on male goats in order to trap and then analyse the stinking volatiles and managed to isolate the most active constituent, 4-ethyloctanal. When female goats smelled this pheromone it triggered ovulation.

You might think that this smell would put them in danger, alerting predators to their presence – but for male goats, the reproductive imperative outweighs the risk. And smelly goats do have ways of avoiding being eaten. In the wild, they prefer to feed as a herd near steep rocks. Lots of eyes are on the lookout and goats' eyesight is excellent, augmented by having horizontal slits for pupils, an adaptation that improves peripheral vision. If one goat spots danger the whole herd quickly climbs to craggy, safer slopes.

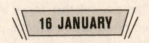
16 JANUARY

Can you create clouds at home?

YES, YOU CAN! All you need is a clear, flexible 2-litre plastic bottle with a screw cap, matches and some water.

Add just enough water to the bottle to cover the base, and shake it around. Light a match, let it burn for a couple of seconds then blow it out. Immediately drop the smoking match into the bottle and screw the lid on quickly and tightly. Now squeeze the bottle hard four or five times. A cloud will appear in the bottle after you have added the match. But when you squeeze the bottle the cloud disappears. Release the bottle and there is the cloud again . . . you've just created a small cloud chamber.

In the bottle water combines with air to form water vapour. This gives rise to the cloud in the bottle. But for water vapour to form this cloud, particulates also need to be present. In this case smoke provides the necessary particles, which act as nucleation sites, allowing drops to collect on them. Without the smoke no cloud will appear.

Squeezing the bottle raises the pressure. This increases the temperature inside the bottle, helping the water to change from visible liquid back to invisible gas (most liquids turn into gases as temperature increases), and the cloud disappears. Releasing the pressure reverses the effect.

Real clouds form in exactly the same way. In this case the water vapour comes from the evaporation of seawater, rivers and lakes. This expands and cools as it rises. There is a limit to the amounts of water vapour the air can hold and this is higher when the air is warmer. As the temperature drops at higher altitudes the vapour begins to condense and clouds are formed, just as in the bottle. And, because the atmosphere contains many small particles, ranging from dust to smoke or salt particles, there are plenty of nuclei around which condensation can gather to form clouds if the pressure and temperature conditions are correct.

17 JANUARY

Where on our planet is the furthest point from any sea?

THE FURTHEST POINT from the sea or, to give its technical name, the continental pole of inaccessibility (CPI), lies in Asia. It is located at 46° 17′ N, 86° 40′ E, in the Dzoosotoyn Elisen in Xinjiang, China, and is 2,648 kilometres from the nearest coastline, at Tianjin on the Yellow Sea.

Due to its unique geographical status, the CPI attracted the attention of Western nuclear strategists debating the relative merits of weapons systems in the 1960s. For proponents of the submarine-launched Polaris missile, the ability to hit any point on the Earth – even if there is nothing there worth hitting – became a key point in the public relations battle with the sponsors of land-based and air-launched weapons.

When the A3 version of Polaris brought the CPI within range in the late 1960s, it was hailed as a technological triumph – particularly by the UK's Ministry of Defence. They did not, however, trumpet the fact that to strike the pole a large nuclear-powered submarine would practically have to visit Tianjin docks. Ironically, by the time Western navies acquired the capacity to bombard all of China with submarine-launched missiles, the region around the CPI was probably featuring more prominently on the targeting lists of generals in Moscow, rather than London or Washington DC, as Xinjiang acquired vital strategic significance in the Sino-Russian confrontation of the last quarter of the 20th century.

The CPI is subject to extreme climatic continentality because it is so far from the moderating influence of the ocean. Indeed, this part of Xinjiang might be considered an extension of the Gobi, which is a decidedly cold desert where temperatures drop to -40 °C in winter. At the other extreme, during daytime in summer it can reach a blistering 50 °C, though temperatures can vary by as much as 32 °C within a 24-hour period. 'Elisen' means 'desert' in the local Chinese Uighur dialect, and although the location is certainly sandy, it is no beach.

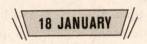

18 JANUARY

Why does a kettle sing?

IF YOU LEAVE the lid off your electric kettle and switch on, you can see what is happening. The heating element quickly becomes covered with small silvery bubbles, each about 1 millimetre in diameter. These are air bubbles, forced out of solution by heat from the element. Rough parts of the element's metal surface provide nuclei for their growth and they eventually detach from the hot element and rise to the surface. These bubbles form and burst silently, and are clearly not the cause of the kettle singing.

After about a minute, the air bubbles are replaced by innumerable smaller bubbles of superheated steam that cling to the growth nuclei on the heating element. A few seconds later, these primary steam bubbles become unstable. As each bubble forms, its buoyancy tends to pull it away from the hot surface. Being surrounded by water which is still far below boiling point, the primary steam bubble suddenly condenses, collapsing implosively. Curiously, the bubble does not vanish completely, but leaves behind a minute secondary bubble, presumably of water vapour, that does not immediately condense but is whirled away by the convection currents. Soon there is such a cloud of these secondary bubbles that the water becomes turbid for half a minute or so.

Meanwhile, the shock waves transmitted through the water by the imploding primary bubbles produce a sizzling sound. You can give this sound a more definite pitch by temporarily replacing the kettle lid. This defines a volume of air above the water surface that resonates to some of the frequencies present in the shock waves.

Soon, the cloud of secondary bubbles clears, and there is a general increase in the size of the primary steam bubbles that are still forming on the element. These are no longer forced to collapse immediately and implosively, since the surrounding water is now practically at boiling point, so the noise fades away. As they grow, streams of buoyant primary bubbles detach themselves from the surface of the element, condensing in the cooler water a centimetre or so above it.

Within seconds, the water becomes hot enough to allow large detached primary bubbles to reach the surface, and now you can hear only the return of sound with the low gurgle of their bursting in the air cavity above the water.

19 JANUARY

Do cats always land on their feet?

THIS QUESTION REMINDS us of a joke. If cats always land on their feet and toast always lands buttered side down, you can construct a perpetual motion machine by simply strapping a slice of buttered toast to a cat's back. When the cat is dropped it will remain suspended and revolve indefinitely due to the opposing forces.

Jokes aside, there might be more to this stereotype than meets the eye. 'High-rise syndrome in cats', a study reported in the Journal of the American Veterinary Medicine Association in 1987 by W. O. Whitney and C. J. Mehlhaff, two New York vets, examined injuries and mortality rates in cats that had been brought to the hospital following falls ranging from between 2 and 32 storeys. Overall mortality rates were low, with 90 per cent of the cats surviving. However, the study unexpectedly found that the incidence of injuries and death peaked for falls of around seven storeys, and then actually decreased for falls from greater heights.

The study was summarised in a *Nature* article, which presented three main variables that determine injury and mortality rate – the speed reached by the moggy, the distance in which said moggy is brought to a stop, and the area of moggy over which the stopping force is spread. While concrete streets work in nobody's favour when it comes to stopping falling items, cats suffer relatively little injury (compared to their owners) because they reach lower terminal velocities and absorb the shock of stopping so much better. A falling cat has a higher surface area to mass ratio than a falling human, and so reaches a terminal velocity of about 100 kilometres per hour (about half that of humans). They are also able to twist themselves so that the impact is spread over four feet, rather than our two. And as they are more flexible than humans, they can land with flexed limbs and dissipate the impact forces through soft tissue.

To answer the paradoxical increase in survival rates once seven storeys has been reached, the authors suggested that an accelerating cat tends to stiffen up, reducing its ability to absorb the impact. However, once terminal velocity is reached, there is no longer any net force acting on the cat, and so it will relax, increasing both its flexibility and the cross-sectional area over which the impact is dissipated once the cat hits the ground.

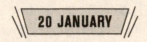

What were the first words?

IT'S A FAIR guess that there was once an original mother tongue – the ancestor to all living and dead human languages. The evidence for this is that all human languages, unlike other forms of animal communication, string together words into sentences that have subjects, verbs and objects ('I kicked the ball'), and anyone can learn any language.

Comparative linguists search for sounds that come up again and again in languages from all over the world. Merritt Ruhlen at Stanford University in California, for example, argues that sounds like *tok*, *tik*, *dik*, and *tak* are repeatedly used in different languages to signify a toe, a digit or the number one. Although studies by Ruhlen and others are contentious, the list of words they say are globally shared because they sound almost the same also includes *who*, *what*, *two* and *water*.

Another approach is to look at words that change very slowly over long periods of time. Words for the numbers 1 to 5, and words involved in social communication, like *who*, *what*, *where*, *why*, *when*, *I*, *you*, *she*, *he* and *it* are some of the slowest evolving. This suggests that language evolved because of its social role.

More broadly, we can say with some confidence that the first words probably fitted into just a few categories. The first ones may have been simple names, like those used by some of our primate relatives. Vervet monkeys give distinct alarm calls for leopards, martial eagles and pythons, and young vervets must learn these. In humans, *mama* is a strong candidate for a very early noun, given how naturally the sound appears in babbling and how dependent babies are on their mothers. The sound '*m*' is also present in nearly all the world's languages.

Imperatives like *look* or *listen* are also likely to have appeared early on, perhaps alongside verbs like *stab* or *trade* that would have helped coordinate hunting or exchanges. Even this simple lexicon allows sentences like 'look, wildebeest' or 'trade arrows'. Finally, simple social words like *you, me* and *I, yes* and *no,* were probably part of our early vocab. Amusingly, a recent study suggested that *huh* is universal, prompting headlines that it was among the first human words. Perhaps it was the second.

21 JANUARY

Could we hear solar wind if Earth's atmosphere extended to the Sun?

We and the sun are in each other's atmospheres already, but most of the gas between us is so tenuous that the sound of the solar wind against our magnetosphere cannot compete with our traffic, pop music and strife. Even if that gas was as dense as our atmosphere at sea level, the loudest solar noises would probably be inaudible infrasonic rumblings, and much weakened after such a journey.

We would have more than noise to worry about, however. The sun would hardly be seen: our atmosphere would be too dense for much light to penetrate if it extended the 150 million kilometres to the sun. If the Earth's atmosphere were to spread as far as the sun and have sufficient density to conduct sound, it would probably be solid ice for most of the distance, causing strange gravitational and orbital effects and burying a human observer on the Earth's surface. Our atmosphere would have a mass of well over 15×10^{30} tonnes – thousands of times greater than the sun plus all its planets. It would collapse and the resulting blast could probably sterilise any planet within several light years.

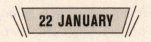

How do people sing well?

A GREAT ITALIAN tenor once gave permission for his larynx to be studied after his death. Air blown through his isolated vocal cords produced the same noise (somewhere between blowing a raspberry and the sound of a whoopee cushion) as other, musically uncelebrated corpses. So why do some people sound like Adele, while others sound like wounded hippos?

Like all musical instruments, human voice quality is largely determined by resonance. Vocal sounds originate in the air stream that is forced through the larynx, the front of which is visible as the Adam's apple. Near the bottom of the larynx are the vocal folds, a pair of flaps that are essentially muscles whose thickness, area, shape and tension can be controlled. The folds are open when you breathe, but when you produce a sound they come together and the air pressure builds up below them until they are forced apart. After reducing the pressure in a puff of air, the folds close again.

After leaving the larynx, the air stream passes through the oropharynx, the upper part of the gullet, and into the mouth. Together, these structures can be likened to the tubing of a brass instrument between the mouthpiece (the vocal folds) and the bell (the lips). Like the air in any tube, it has resonant frequencies, known as formants. Changing the shape of the tube by arching the tongue, opening the jaws, modifying the shape of the lips or altering the position of the larynx will either raise or lower the frequency of each formant. To a certain extent we do this unconsciously, but singers learn how to control these parts of their anatomy.

Singers also have another formant, which is thought to be caused by a standing wave set up in the short tube between the vocal folds and the point where the relatively narrow larynx joins the wider oropharynx. In acoustical terms there is an impedance mismatch at that junction so part of the sound energy is reflected back towards the vocal folds. The effect is weak in normal speech because the tube is very short, and in untrained singers it can shorten still further because their larynx tends to rise as they try to sing louder and at a higher pitch. But in trained singers the larynx descends, lengthening the tube, adding clarity and projecting the voice. For these reasons it is called the singer's formant.

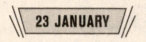

23 JANUARY

Why do three buses always turn up at once?

It's a popular cliché, but there is, in fact, a technical term for several buses arriving at once: it's known as 'bunching'.

If for some reason a bus is delayed by a few minutes, there will probably be more people waiting for it than on average. This is especially true when the frequency of buses on a particular route is high enough (one every 10 minutes, say) that passengers tend to arrive at stops randomly rather than according to the timetable.

Any late-running will therefore increase the time a bus has to pause to pick up passengers at a stop, especially if the bus driver has to sell or validate tickets on entry. The late bus is therefore made slightly later still. This effect is compounded at each stop, causing more and more people to be waiting, delaying it even more.

Meanwhile, the next bus on that route is getting a pretty quick run because many of the passengers it would have picked up are on the late-running bus. Eventually it catches up the bus in front and, if it doesn't overtake it, we are left with two buses trundling along together, with the next service behind catching up on them too.

The most obvious and widely practised solution is to include 'timing points' along the route – stops where a bus is scheduled to wait for a few minutes before continuing. A late-running bus may ignore this wait and so make up a few minutes, but on the downside a bus that is on time will put passengers through an unnecessary delay whenever it reaches a timing point.

24 JANUARY

What would happen if the Moon disappeared?

THE MOST IMMEDIATE difference would be the disappearance of the tides. Both the sun and moon influence the tides on Earth, but the moon is the dominant force. Remove the moon and the daily rush of the tides would recede to a gentle ripple.

The next omen of doom would be wild swings in the Earth's rotational axis from a position almost perpendicular to the ecliptic plane all the way to being practically parallel to it. These swings would provoke drastic climate changes: when the axis points straight up, each point on the globe would receive a constant amount of heat throughout the year but, when the axis lies parallel to the ecliptic, Earthlings would spend six months of the year sweltering under the unending blaze of the sun, only to spin round and shiver for the next six months, hidden on the frigid surface of the Earth's dark side.

Of all calamities, though, the creature to be pitied first is the marine organism called 'nautilus'. This mollusc lives in an elegant shell shaped like a perfect spiral partitioned off into compartments. The nautilus only lives in the outermost partition, and each day adds a new layer to its shell. At the end of each month, when the moon has completed one revolution around Earth, the nautilus abandons its current compartment, closes it up with a partition, and moves into a new one. Scientists have proved that the number of layers making up a chamber are directly linked to the number of days it takes the moon to circle the Earth. Remove the moon and the nautilus lies stranded, forever locked in the same chamber and wishing ruefully for the days when it could look forward to a new home.

Thankfully, it seems unlikely that an alien spaceship would come along and steal the Moon, even for civilisations with a sense of humour that is advanced many millions of years beyond our own.

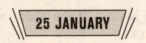

Why do people have eyebrows?

EYEBROWS SERVE A functional purpose in keeping rain or sweat from your forehead out of your eyes. But perhaps more importantly, we use our exceptionally mobile eyebrows to communicate our emotions.

The position of the eyebrows emphasises expressions on the human face thus giving others an accurate picture of the individual's mood. This gives a good indication of whether a person is friendly or whether they might be dangerous to approach. Perhaps most important is the 'eyebrow flash', a rapid up-and-down flick of the eyebrows that conveys recognition and approval. The ability to telegraph friendly intentions from a safe distance would have had obvious survival value for our ancestors. Eyebrow signalling of various kinds is widespread among primates, although only in humans are the eyebrows highlighted by setting them against bare skin

Smiles come in many forms, from expressions of merriment or contentment to leers, smirks and even anger. The position of the brow, emphasised by the eyebrows, is what gives us a visual cue to what an individual is really feeling.

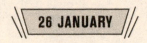

Why are there no three-legged animals?

A TRIPOD IS wonderfully stable, so there could be something to be said for having three legs. When insects walk, they use their legs as two sets of three. At any instant their weight is supported by three legs – two on one side of the body and one on the other. Meanwhile, the other three legs can be moved forward to form the next 'tripod'.

Kangaroos have strong tails capable of bearing weight, and though they do not have any 'three-legged' gaits, they can move slowly with a 'five-legged' gait. First the tail and forelegs are used to support the animal while the hind legs are brought forward in unison, then the hind legs take the weight while the kangaroo shifts forward before putting its forelegs and tail back onto the ground. Because the forelegs are short, the head stays close to the ground throughout, making this gait good for grazing.

The first vertebrates to walk evolved from fish, which swim with a lateral motion, so the gait they evolved probably also involved side-to-side movement. If fish had evolved differently, swimming with a vertical tail motion like a dolphin, then the first vertebrates would have had a gait with some up-and-down motion, possibly using the tail as a 'leg'. In this alternate reality, a five-legged gait similar to a grazing kangaroo could have been common, and tripedal creatures could conceivably have evolved.

Most animals are bilaterally symmetrical, however, so it is not surprising that their limbs come in pairs – two in the case of land-dwelling mammals, three in insects, four in spiders, and various larger numbers in crustaceans, centipedes and millipedes. Three legs is an unlikely arrangement which doesn't confer an advantage in movement over two or four.

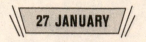

What causes different types of rain?

SOMETIMES RAIN FALLS down in great elongated stair rods, other times there's just a misty drizzle. In Afrikaans (and in Welsh) when heavy rain lands, they're called 'old women with clubs', the circular sheet of splashing water suggests a wide skirt and the centrally rebounding droplet a cudgel.

Rainfall intensity depends mainly on the depth of the cloud and the strength of the updraughts. Rapidly rising air produces fast condensation of water droplets and large amounts of rain, mostly when the cloud extends high enough for ice crystals to form among supercooled water droplets.

Shallow clouds with weak updraughts only give drizzle, which rarely falls faster than 3 metres per second. Large raindrops can reach a terminal velocity of about 10 metres per second. Their fall speed increases with size until the diameter approaches 6 millimetres, at which point wind resistance flattens the base, increases the drag and prevents further acceleration.

However, if the rain is caught in a 'downburst' where an air column is descending at 20 metres per second or more, the rain hits the ground harder. Downbursts are often associated with cumulonimbus clouds that contain almost vertical air currents. The weight of precipitation in the cloud may be enough to trigger a downburst.

Rain from deep flat cloud layers is usually caused by slow diagonal ascent along a sloping frontal surface. Such rain is persistent but seldom heavy. This can change if prolonged lifting makes the layer unstable. Then massive turrets containing strong updraughts grow vertically out of the layer. These can produce heavy downpours from a cloud mass which had previously only given moderate rain.

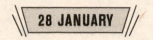

How much does a human head weigh?

MEASURING THE WEIGHT of your head involves effectively isolating it from the rest of your body. But while decapitation is an obvious method it has far too many drawbacks, not least that you will not be able to see the results of your experiment.

Happily there is an easier way and it is surprisingly accurate. You will need a bucket of water filled to the brim, a larger empty vessel big enough to accommodate the bucket and its contents, and your head. Because it involves immersing your whole head in water, make sure that there are always at least two people present and that children are supervised.

Place the full bucket inside the larger, empty vessel. Take a deep breath and lower your head, crown downwards, into the bucket until the water reaches the base of your chin. Hold your head still for as long as possible until any water ripples cease, then take it out and have a few gulps of air.

Collect all the water that has spilt into the outer vessel and measure its volume. If you don't have a larger vessel to catch the water, let all the displaced water run away and then start to refill the bucket into which your head was placed with a measuring jug, taking a note of exactly how much water you need to refill the bucket to the brim.

Although bone and brain are constituent parts of our heads, the rest, like the remainder of our bodies, is mostly water. We know that 1 litre of water weighs 1 kg, so if we can measure the volume of water displaced by a head we can approximately work out how much that head weighs. If you displace 4 litres of water using the experiment above, your head weighs approximately 4 kg, roughly the weight of the average head.

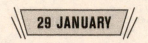

Will we ever speak dolphin?

DOLPHINS AND HUMANS can communicate, but will it ever be possible for them to engage in meaningful conversation? Perhaps, but communication between the two species has been limited to date.

The idea that dolphins possess a communication system as sophisticated as human language was proposed by John Lilly in the 1960s, who vowed that pioneering researchers would one day 'crack the dolphin code' and begin an interspecies dialogue. In the ensuing years, dolphins were taught to use artificial symbol systems, with equivocal results. Their performance is comparable to great apes where comprehension is concerned, but when it comes to using symbols to establish two-way communication with humans, dolphins have been overshadowed by linguistic prodigies such as Kanzi the bonobo.

Dolphins' own communication system has been the subject of much study, revealing a perplexing array of vocal and non-vocal signals. But a sober view of half a century's worth of evidence suggests that dolphin communication – even when taking into account the referential, or word-like, nature of their mysterious 'signature whistle' – is nothing more than a variation on the type of communication system seen throughout the animal world. It is complex, to be sure, though likely to be short on content.

Science is destined to make great strides in unravelling the mysteries of dolphin communication, as there is much we do not yet understand about the function of their vocalisations. However, the idea that dolphins are harbouring a secret language that awaits decryption is looking increasingly like a spot of wishful thinking from a bygone era.

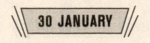

Why do some flowers close at night?

WHEN FLOWERS CLOSE temporarily for the night they are effectively in standby mode, protecting their delicate reproductive parts and pollen while they are not in use. The pollen is isolated from the dew that forms during the night, keeping it dry so that it can be dusted onto a passing insect the following day. Indeed, some flowers remain closed until some time after dawn, and only reopen when the day is warm enough for the dew to have evaporated.

Closing the flowers also helps to protect against nighttime cold and bad weather. As well as closing their petals, some plants also close the tough surrounding structures, called bracts, to protect the flower against plant-eating insects. Keeping the pollen dry while limiting access for plant-eating insects – and the fungi and bacteria that they carry – also means that the pollen is less likely to spoil.

Ultimately all these adaptations minimise wastage of pollen or damage to the flower. Analogously, many moth-pollinated flowers release their fragrance only at night, so avoiding waste during daylight.

Some flowers open and close so punctually that a once-popular gardening fashion was to plant flowers in sectors of a bed resembling a clock face. These were so planned that flowers opening in each sector matched the position of a notional hour hand on the clock. In season, all being well, one might actually be able to tell the time by consulting one's flowering clock.

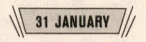

What colour is a sub-atomic particle?

WHEN WE LOOK at things, we see light coming from their surfaces. For most everyday objects we see reflected light. If the light falling on the object is white, then the colour comes from the parts of the visible spectrum not absorbed by the surface. So a 'blue' object is one that has absorbed all the other colours, leaving only the blue component to be reflected.

But light can also be emitted by the object itself. All objects at temperatures above absolute zero are constantly emitting electromagnetic radiation. At room temperatures, this light is at a wavelength below the visible spectrum so we cannot see it with our naked eye (although we can see it using infrared goggles). However, when an object becomes hot enough, its electromagnetic radiation enters the visible spectrum. It starts at the lower red end (for example, a hot glowing coal), then more colours are added as the object gets hotter, until eventually it glows 'white hot'.

If you have an isolated subatomic particle such as a lone electron, proton or neutron, then light waves in the visible spectrum pass right by it without interaction because the particle is so small. So there would be no reflected light to give the particle any colour you could see. It would be invisible, just like the oxygen and nitrogen atoms in the air you are looking through to see this page.

The only way to see such a subatomic particle with the naked eye would be if it were hot enough to emit electromagnetic radiation in the visible spectrum – so its colour would depend on the temperature of the particle. There is an everyday example of this. Our sun is a ball of plasma made up mostly of free protons and electrons that emit a white light. Other stars have lower surface temperatures, so their protons and electrons look red, while other stars are hotter and have a bluish tinge.

FEBRUARY

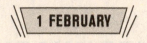

1 FEBRUARY

What is MSG?

FOR THOUSANDS OF years the Japanese have incorporated a type of seaweed known as kombu in their cooking to make food taste better. It was not until 1908, however, that the actual ingredient in kombu responsible for improvement in flavour was identified as glutamate. Today, hundreds of thousands of tonnes of monosodium glutamate, or MSG, are produced all over the world.

Monosodium glutamate contains 78.2 per cent glutamate, 12.2 per cent sodium and 9.6 per cent water. Glutamate, or free glutamic acid, is an amino acid that can be found naturally in protein-containing foods such as meat, vegetables, poultry and milk. Roquefort and Parmesan cheese contain a lot of it. The glutamate in commercially produced MSG, however, is different from that found in plants and animals. Natural glutamate consists solely of L-glutamic acid, whereas the artificial variety contains L-glutamic acid plus D-glutamic acid, pyroglutamic acid and other chemicals.

It is widely known that Chinese and Japanese food contains MSG, but people don't seem to be aware that it is also used in foods in other parts of the world. In Italy, for example, it is used in pizzas and lasagne; in the US it is used in chowders and stews, and in Britain it can be found in snack foods such as potato crisps and cereals.

It is thought that MSG intensifies the naturally occurring 'fifth taste' in some food – the other, better known, four tastes being sweet, sour, bitter and salt. This fifth taste is known as umami in Japanese, and is often described as a savoury, broth-like or meaty taste.

Umami was first identified as a taste in 1908 by Kikunae Ikeda of the Tokyo Imperial University, at the same time that glutamate was discovered in kombu. It makes good evolutionary sense that we should have the ability to taste glutamate, because it is the most abundant amino acid found in natural foods.

John Prescott, associate professor at the Sensory Research Science Center at the University of Chicago, suggests that umami signals the presence of protein in food, just as sweetness indicates energy-giving carbohydrates, bitterness alerts us to toxins, saltiness to a need for minerals and sourness to spoilage. A team of scientists has even identified a receptor for umami, which is a modified form of a molecule known as mGluR4.

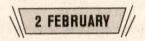

2 FEBRUARY

Why do men have nipples?

MANY SUGGESTIONS FOR this phenomenon have been offered. They may exist to help men check that their vests are on straight, or be present as a safety feature – to warn us how far out from the beach we can safely wade.

However, there is a more plausible explanation. Male and female human embryos are identical in the early stages of their development. If the fetus receives a Y chromosome from its father, a hormonal signal is produced: the labia fuse to form a scrotum, the gonads develop as testicles and a male results. Otherwise the 'default' female remains.

Various structures in the adult reflect the symmetry of male and female and their common embryonic source. Men have nipples because they have already begun to develop when the 'switch to male' signal is received. The development of breasts is halted in most – but not all – cases, but the nipples are not reabsorbed.

Another effect of these developmental pathways which are shared by both males and females is pointed out in Stephen Jay Gould's essay 'Male Nipples and Clitoral Ripples'. Males need plenty of blood vessels and nerve endings in their penises to achieve erections. Because the penis and clitoris have their origins in the same structure, females have the same number of blood vessels and nerve endings packed into a much smaller area, resulting in the enhanced sensitivity of the clitoris.

Conclusive evidence that God is not a man?

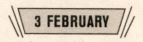

3 FEBRUARY

How do weather forecasts know when it's going to rain?

A CRUCIAL PIECE of information needed to produce an accurate weather forecast is knowing exactly how much water is contained in a cloud. In the UK, the Council for the Central Laboratory of the Research Councils' Chilbolton Observatory uses a Doppler radar to find out.

The choice of frequency for the radar beam is very important. If the beam interacts too strongly with the water in a cloud, in terms of either reflecting or attenuating the signal, then the radar will have only a limited ability to penetrate cloud structure. If the interaction with water is too weak, then no useful information can be returned at all.

The Chilbolton facility can analyse and extract a huge variety of data and it has a maximum range of around 160 kilometres. It is able to provide information on droplet density, size, speed, and whether the droplets are water or ice.

Using this tool, you could work out fairly accurately the total water content contained within a cloud and, from the structure of the cloud, how likely it is to start raining – this technique has proved very useful at the Wimbledon tennis championships in past years, which are, of course, notorious for being interrupted by downpours.

Such radars help to produce detailed information on weather, from tracking hurricanes to helping to produce your daily weather forecast and predicting areas of turbulence on aircraft flights.

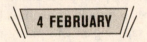

What would a DNA-free meal look like?

ALL LIVING ORGANISMS have DNA, so you'd have a hard time finding something to eat that didn't contain any tissues or cell cultures. You could try eating RNA viruses, but you'd need to produce them in a cell culture, which generally requires animal serum to keep the cells alive.

One menu option is red blood cells. In many species, including humans, the nucleus and mitochondria are removed from these cells during the maturation process. This is to make room for more haemoglobin, the iron-bound protein that carries oxygen. Because the nucleus and mitochondria contain all the cell's DNA, you could argue that provided you don't kill the animals, drinking their blood is the ultimate vegetarian diet. You'd need to filter out the white blood cells, which still have plenty of DNA, but the rest of the blood components would be fine. They'd provide you with protein, some sugars and vitamins, but probably more iron than is healthy.

If that doesn't sound appealing, consider totally (bio) synthetic foods. Biologists routinely construct yeast and bacterial lines designed to churn out large quantities of a specific protein or other biological molecule. Assuming it would be possible to scale this production up, you could produce sufficient quantities of purified proteins, sugars and so on to act as a food source. Don't expect it to be tasty, though: the proteins and sugars produced would be purified from the culture as crystalline powders. The result would either be oil or a pretty nasty goo.

Many, perhaps all, of the various vitamins and other nutrients we require could probably be synthesised in similar ways, given time and cash. The various mineral compounds we need – iron, copper, zinc, iodine and so on – are probably available from a good synthetic chemist. And, of course, you could drink milk. It's a complex mixture of secreted proteins, fats, sugars and pretty much everything else you need to stay alive. It may contain cells from the animal which produced it, but you could probably centrifuge these out.

Finally, if you've got the budget: one cubic metre of lunar soil contains enough of the right elements to make a cheeseburger, an order of fries and a fizzy drink. That would contain no DNA, but might be a little expensive.

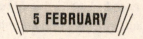

5 FEBRUARY

What would happen if you jumped into a pool of jelly?

JELLY IS AN interesting substance because it behaves as both a solid and a liquid.

The active ingredient in jelly dessert (or jello) is gelatin, a protein-based gelling product made from collagen. Gelatin comes in different grades, or Bloom numbers, as measured by the force required to push a plunger into a solution of the stuff to a predetermined depth: the more rigid the sample, the higher the Bloom number. Jelly babies – a popular British sweet shaped like a miniature baby – have a high Bloom number, so there is little danger of drowning in a pool of the mixture used to make them.

The density of jelly is typically 10 per cent higher than water, so a swimmer would float higher in a pool full of jelly than in a pool full of water. Jelly is also more viscous than water, meaning that someone diving into jelly might have difficulty surfacing. However, two researchers from the University of Minnesota, Minneapolis, won the 2005 Ig Nobel prize for chemistry for showing that people could swim just as quickly in water spiked with guar gum, an edible thickening agent, as in ordinary water. The spiked liquid has double the viscosity of water, yet the increased drag is cancelled out by the increase in thrust that swimmers can generate in it.

A person jumping into a pool of jelly would be slowed dramatically in the first fraction of a second – doing a belly-flop onto jelly would hurt. They would then sink gradually until their buoyancy became neutral, although they would float higher than in water as jelly is a little denser.

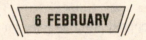

6 FEBRUARY

How do birds recognise each other?

MANY BIRDS USE acoustic recognition and can identify each other's voices. Swallows, finches, budgies, gulls, flamingos, terns, penguins and other birds that live in larger flocks do this. Odours can also play a role in determining how some birds recognise each other.

In ducks, sound seems to be the principal method of recognition: they have been fooled into returning to the wrong nest, only to be greeted by a portable cassette player rather than their ducklings. The ability to recognise their own young saves colony-living birds from expending energy in raising someone else's offspring. It also stops ducklings running the risk of aggression from adults if they beg food from the wrong ones. Natural selection favours individuals who know who they are talking to.

Chasing away non-descendant young is called 'kin discrimination'. Waterfowl have long been thought to be unable to keep track of their own young. They have been seen to lose their own ducklings to another parent, or to mistakenly accept and care for non-descendant ducklings. This has been put down to the fact that birds do not generally have a central family unit.

Ducks do behave in a different way towards their own ducklings, though. Parents sometimes favour their own offspring over non-descendant young, or they may tolerate or encourage the ducklings to mix. Consequently, some provide what is called alloparental care, a form of adoption. This is seen when a duck is able to increase the chances of survival of her own offspring by accepting non-descendant ducklings into her entourage. Her own ducklings might be better off because the risk of any individual being eaten by a predator is lower if it is part of a bigger group. To improve the advantage even more, the non-descendant ducklings may be positioned at the edge of the brood, further away from parents. This has been seen in Canada geese; the adopted goslings were noted to generally potter further away from their adoptive parents than the biological offspring, and therefore not as many survived.

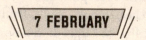

7 FEBRUARY

How do flowers drink water?

If you forget flowers in a vase, you'll discover the astonishing rate at which they take up water. A full vase can become empty in no time at all, leaving you with drooping stems and fallen petals.

Plants need water in order to live and grow, and they do this by drawing up water from the soil through their roots. Normally all this is invisible, but adding some food colouring to a vase of white flowers can show you exactly what is going on.

The coloured water is drawn through vascular structures (the xylem) in the plant. At its simplest level xylem acts like a straw. As water is transpired – or 'sucked' – away from the petal or leaf surface, more water is pulled into the straw to replace it. Xylem conduct water and dissolved nutrients upwards from the root. At their most dense, the hollow channels of xylem form the internal scaffolding of woody stems as they branch out through leaves and petals. You can see the structure of xylem if you remove one of the flowers from the vase and cut through the stem. Vein-like tubes will be clearly visible in the cross-section, filled with coloured liquid.

Over the course of about 10 hours you'll see the spread of the coloured water throughout your flowers, until they are entirely dyed. Plants, especially in hot conditions, can take up vast amounts of water, which is why a vase that was full one morning can be almost empty the next. On a hot day, a full-grown tree, such as a birch or sycamore, can draw upwards of 500 litres from the ground, in some cases aided by root pressure, forcing water upwards into the xylem; all of which is later transpired through evaporation.

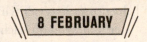

8 FEBRUARY

Why do tunes get stuck in your head?

GETTING A SONG stuck in your head is known by many different names including stuck song syndrome, or earworm, a term translated from the German word *Ohrwurm*.

Daniel Levitin, a neuroscientist at McGill University in Montreal,

Canada, has suggested that it has an evolutionary origin. Before writing was invented, just 5,000 years ago, songs helped people to remember and share information. Levitin suggests that variations of rhythm and melody provide the cues for easier recall, something that continues in communities with strong oral traditions.

This chimes with the findings of James Kellaris, professor of marketing at the University of Cincinnati, Ohio, who says earworms occur when you subconsciously detect something unusual about a piece of music. Usually between 15 and 30 seconds long, an earworm is replayed in your mind in a loop and is difficult to dislodge.

Music that is repetitive and simple, yet with an unexpected variation in rhythm, is most likely to become an earworm. For example, the repetitive melody and shifting time signatures of the song 'America' from *West Side Story*. Kellaris claims that 98 per cent of people experience the feeling. About 74 per cent of earworms are songs with lyrics, 15 per cent are jingles from adverts, whereas instrumental pieces account for only 11 per cent of earworms.

Victoria Williamson, a music psychologist at Goldsmiths, University of London, suggests a number of triggers. Earworms are more likely to be bits of songs that have been heard recently or repeatedly. A song associated with a stressful or stimulating experience is also a good candidate.

In the early 1980s, Myron Warshauer tried to exploit this by patenting a 'musical floor reminder system' in multi-storey car parks in the US. The system helped people recall which floor they parked on by associating music and murals with each one.

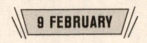

Are Arabic numerals Arabic?

ALTHOUGH THE NUMERALS used by Europeans, Americans, and much of the international community are referred to as Arabic numerals they were not originally created by the Arabs. This misnomer originated in the 9th century when a manuscript on arithmetic which had been written in India was translated soon after into Arabic. Merchants then carried this book to Europe where it was subsequently translated into Latin.

Because the source for the Latin translation was an Arabic text, the numerals were falsely ascribed to the Arabs. This is where the confusion arose: the numerals came from India and are not Arabic at all.

India made two crucial advances in number systems. They introduced place values and a special symbol for zero. The value of these innovations can readily be experienced by anyone trying to do simple arithmetic with non-place-value systems, like Roman numerals. Place-value numeration allows the mathematician to arrange addition in columns and to carry over numbers from the ones to the tens, or from the tens to the hundreds columns. This is very difficult to do with Roman numerals.

The present set of Arabic numerals evolved over the early centuries but has changed little since the advent of printed books in around 1445. Curiously the numerals 4, 5, 6 and 7 underwent the most changes from the original script. Prior to the printing press, the glyph we now use for four would have been considered to represent five.

It is not surprising that the actual glyphs used in Arabic-speaking countries are very different from those that we use now, and these differences can be very confusing, especially to the unwary.

The traffic speed limit on the bridge running between Khartoum and Omdurman in Sudan used to be posted in both English and Arabic. The limit was 10 miles per hour, but apprehended drivers frequently defended themselves on the grounds that they had never exceeded the 15 miles per hour that they had seen on the signpost. This is because the glyph we use for zero is very similar to that used in Arabic for five.

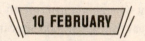

10 FEBRUARY

Can I trick my sense of touch?

TAP YOUR NOSE with your finger. Did you feel one touch or two?

Most of us will feel only one touch, likely from your fingertip. This seems counterintuitive – the sensation from nerves in your nose has only a few centimetres to travel to your brain, while the one from your fingertip has to travel about a metre up your arm and shoulder. What's going on?

Your brain does register two touches – the touch of your finger on your nose, and the touch of your nose on your finger – but the illusion arises because you have many more sensory receptors on your fingertips than on your nose, and they are more sensitive in different ways. As an aside, look up 'somatosensory homunculus' on the internet.

If you stub your toe, the first thing you do is grab hold of the injured part with your fingers to see if it's OK. You use your fingers because they provide better information about that part of your body than the body part itself.

So you do feel two sensations. Your brain just chooses to disregard the sensory stimulation of the skin on your nose because it is a much less rich source of information.

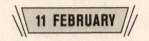
Where did Earth's oxygen come from?

PRACTICALLY ALL OF the atmospheric oxygen is of biological origin. The main culprit, however, is not plants but humble cyanobacteria. These single-cell organisms, which were present on Earth more than 3.5 billion years ago and pre-date plants, were initially responsible for all oxygen production and are still responsible for more than 60 per cent of current oxygen production.

Cyanobacteria come in many varieties and are sometimes called blue-green algae, although they are not really algae. A species of cyanobacteria present in the ocean, *Prochlorococcus marinus*, is both the smallest photosynthetic organism known and the most abundant of any photosynthetic species on the planet. It was only discovered in 1988.

In Earth's crust, oxygen combines with all the most common atoms to form water, rock, organic compounds and almost everything around us. Spontaneous free oxygen is about as likely as finding round rocks perched on steep slopes. Such rocks would imply that something had pushed them uphill more strongly than they could roll downhill.

Similarly, any free oxygen around us has been torn from its compounds with more than its bonding force. And that is a lot of force that only a few things are able to produce. Ionising radiation, such as X-rays, can do it, but there is little of that about. Visible light does it laboriously, step-by-step through photosynthesis, the only process that could release the breathtaking amount (no pun intended) of oxygen that we see about us.

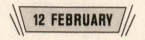

Why does red cabbage make fried eggs green?

CHEF HESTON BLUMENTHAL, one of the founding fathers of molecular gastronomy, would love this experiment. So would Dr Seuss.

Squeeze juice from cooled cooked red cabbage into a jug. Heat oil in a frying pan, and begin to fry an egg until the white is just turning from clear to white. Drip a small amount of cabbage juice into the setting egg white. Where the juice hits, the egg white will turn a lurid green.

This is because red cabbage juice is a good indicator of whether a substance is an alkali or an acid. If added to an alkali, such as ammonia, it will turn green; if added to an acid, such as lemon juice, it will turn red. In neutral substances it is purple, the natural colour of red cabbage. Because egg white (mostly the protein albumen) is alkaline, it turns green. Any number of substances can be tested in this way, although take care to avoid strongly corrosive chemicals such as drain cleaner or bleach because these can be dangerous.

The experiment works because red cabbage contains water-soluble pigments called anthocyanins (also found in plums, apple skins and grapes). These change colour depending on whether they are in the presence of an acid or an alkali. These change the number of hydrogen ions attached to the molecule – acids donate hydrogen ions while alkalis remove them – and it is the presence or absence of hydrogen ions that is responsible for the different colours. This explains why red cabbage that is pickled turns red, rather than its natural purple colour. Pickling takes place in vinegar, which is acidic.

Red cabbage juice breaks down quite quickly, so if you are going to use it to test the acidity or alkalinity of other house-hold foods or products use it sparingly and fast.

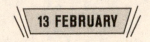

How long a line could a single pencil draw?

OUR MINDS BOGGLED at the range of variables implied by the question, but one reader, Andrew Fogg, decided to conduct a simple experiment in which the variables were reduced to a manageable number.

Taking the simple case of a clutch pencil, he found by experiment that a 1-millimetre length of 0.5-millimetre 2B lead would draw about 9 metres of uniform line on ordinary photocopier paper. In his clutch pencil a new lead has a usable length of 50 millimetres – that's 450 metres of line per lead. Looking at it another way, 1 cubic millimetre of pencil lead is needed to draw 45.84 metres of line.

A brand-new wooden pencil from a reputable maker is 175 millimetres long with a lead diameter of 2 millimetres. Assuming it is possible to use all but the last 20 millimetres of the lead, and (crudely) that each millimetre of lead draws 9 metres of line as with the clutch pencil, that would give us 1,395 metres of line for the whole pencil.

However, the volume of usable lead in the pencil, assuming again that the last 20 millimetres can't be used and that half is lost to sharpening, is 243.5 cubic millimetres. At the same volumetric wear rate as in the clutch pencil, that should produce 11,162 metres of line. The actual output will likely be somewhere between these two answers.

The hardness of the lead will make a difference, as will paper type, the density of the line and how careful the user is not to sharpen too often or too far. Forget string – we now have a new saying, 'How long is a pencil line?'

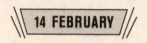

Why do we feel a 'broken heart'?

A METAPHOR OFFERS a clue: the Japanese emphasise the stomach rather than the heart. Also, in English we speak of 'not having the stomach' for something. In healthy people it is the muscular viscera that draw attention to themselves, particularly those of the heart, oesophagus and stomach. Though they are under involuntary nervous and hormonal control, their physical reactions give dramatic feedback.

The heart reflects emotions by the intensity and rhythm of its actions, for example in shock it leaps and pounds. Anxiety can cause actual stomach aches and 'lump-in-throat' oesophageal spasms.

Helplessness causes diastolic flaccidity – a drastic reduction in blood pressure – which may explain deaths in those who believe they are the subject of a voodoo curse. It could also explain the physical heartache of grief, loss or betrayal. In addition, it reduces circulation and causes cardiac irregularity or palpitations, with frightening symptoms such as faintness and tingling in the face and extremities.

Conversely, surges of adrenalin cause the pulse to race, which increases blood pressure for emergency exertion but can cause paralysing panic when one does not know what to do. Compare this to the less muscular vital organs which cannot give such immediate feedback – it takes time to interpret what the liver or kidneys have to say.

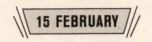

What was the first life on Earth?

IN THE BEGINNING was Ida, the initial Darwinian ancestor – the first material on Earth to transform from inert to, well, ert. Ida begat Luca, the last universal common ancestor, a molecule that stored information as genetic code, and gave rise to all life on Earth.

Ida and Luca live on within us. Our cells all use the same genetic code embodied in DNA, suggesting Luca was itself made of DNA. Except it isn't that simple. All life uses proteins to make DNA and execute its code – but proteins themselves are made from DNA templates. Which came first?

Probably neither. RNA is a close relative of DNA found in all living cells that also carries genetic code and, crucially, can catalyse chemical reactions on its own. The RNA world hypothesis says Luca was born out of an RNA soup that eventually gave rise to DNA and the first cells.

But where did RNA come from? In the 1950s, American chemists Stanley Miller and Harold Urey famously zapped a mixture of gases and water with electricity and ended up with a handful of biotic molecules. Nowadays, though, more nuanced ideas are in vogue. Nick Lane of University College London, for example, thinks that warm vents on the ocean floor provided a soup of methane, minerals and water from which RNA could form. Michael Yarus of the University of Colorado in Boulder, meanwhile, favours a slushy pond whose continual freezing and thawing pushed chemicals together in just the right way.

Intriguingly, more recent experiments trying to coax RNA into existence have shown that when the chemistry is just right, many of life's building blocks seem to form almost spontaneously – increasing the likelihood it happened elsewhere, too.

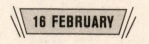

How do paper planes fly?

EVEN A STONE has some very limited aerodynamic features if thrown well – consider how far a shot-putter can heave a metal ball. The mechanism by which a paper aeroplane flies can be simply accounted for by Newton's second law: force (in this case, lift) equals the rate of change of momentum. In simpler terms, the plane flies faster and for longer if you throw it harder unless there is something (such as a flapping piece of paper on the nose or dangling, floppy wings) that causes drag and slows down the plane, stymieing its forward motion.

A sheet of paper cannot be thrown as far as a small stone of the same weight because of drag, which is a combination of air resistance and turbulence. Air resistance, caused by the viscosity of air, is proportional to an object's frontal cross-sectional area – or how big the plane's nose is when you look at it front-on. Turbulence is a result of twisting air currents and vortices that form around the plane. It is proportional to surface area and is reduced by a stream-lined shape.

The ideal shape is long and thin, tapered towards the front and rear like an arrow, with a small surface area for its given weight. Folding a sheet of paper into this general shape will increase the distance it can be thrown. Short wings work better – long ones add weight and drag.

The other key factor to consider is the angle at which the plane sits as it moves forwards. Air needs to strike the underside of the wings and be deflected downwards, ensuring a corresponding upwards force on the plane. If it does not sit at an angle to the flight direction with its nose pointing slightly upwards, it will not generate lift. In a standard paper dart, most of the weight is at the back, which means the back drops, the nose rises and we have lift. This lift balances the weight and the paper dart flies. That's why – unless this lift is too effective, forcing the plane to shoot rapidly upwards and stall – paper darts are better fliers than paper aircraft that actually look like real planes.

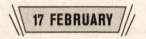

17 FEBRUARY

What does an indestructible sandwich taste like?

FIRST CAME THE atom bomb, the stealth bomber and the airborne laser. Then in 2002 came one of the US military's most fearsome weapons: the indestructible sandwich.

Capable of surviving airdrops, rough handling and extreme climates, and just about anything except a GI's jaws, the pocket sandwich was designed to stay 'fresh' for up to three years at 26 °C (about the temperature of a warm summer's day), or for six months at 38 °C (just over body temperature).

For years the US army had wanted to supplement its standard battlefield rations, called 'Meal, Ready-to-Eat' (MRE), with something that could be eaten on the move. Although MREs already contained ingredients that could be made into sandwiches, these had to be pasteurised and stored in separate pouches, and the soldiers needed to make the sandwiches themselves.

'The water activities of the different sandwich components need to complement each other,' explained Michelle Richardson, project officer at the US Army Soldier Systems Center in Natick, Massachusetts. 'If the water activity of the meat is too high you might get soggy bread.'

To tackle the problem, researchers at Natick used fillings such as pepperoni and chicken, to which they added substances called humec-tants, which stop water leaking out. The humectants not only prevented water from the fillings soaking into the bread, but also limited the amount of moisture available for bacterial growth. The sandwiches were then sealed, without pasteurisation, in laminated plastic pouches that also included sachets of oxygen-scavenging chemicals. A lack of oxygen helped prevent the growth of yeast, mould and bacteria.

Soldiers who tried the pepperoni and barbecue-chicken pocket sandwiches found them 'acceptable'.

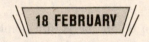

18 FEBRUARY

Why do we have ten fingers and ten toes?

HUMANS' REMOTE TASSEL-FINNED or fringe-finned (crossopterygian) ancestors emerged from the water with a limb architecture of one bone from the shoulder or hip, two bones from the elbow or knee, and several bones from the wrist or ankle. All land vertebrates have limbs that are based on that original scheme, including humans.

These pioneers had lots of slender 'toes' on all four feet – too many and too slender for control and power. There must have been strong selection for a more definite joint structure and more strength in each digit. By the time the first true amphibians appeared, toes had thickened and been reduced to eight or so on each foot.

Long before the first reptiles evolved, five toes had become pretty much standard issue. Mammals continued the pattern, which seems to be so robust and versatile that it has persisted among most non-specialised groups and a good few specialists as well, such as tree climbers and their descendants, including humans.

Specialisation tends to reduce the number of toes. Creatures that run, for example, need light feet more than they need versatile bone architecture, so their toes reduce in number – down to one in horses, and in size – two main toes and a couple of vestigial ones in artiodactyls (cattle, deer and suchlike). Some creatures, such as snakes, have even lost entire limbs.

19 FEBRUARY

Why are some people left-handed and others right-handed?

THE SIMPLE ANSWER is that they've inherited genes for left or right-handedness, which is why handedness runs in families and identical twins are more likely to have the same handedness than dizygotic (fraternal) twins. The genes involved are a little strange, because while one makes people right-handed, the other only makes it random as to whether an individual is right or left-handed. So identical twins with the latter gene can have different handedness.

Genes are only the immediate cause of handedness. Very occasionally, 'biological noise' during development, or brain or arm trauma, will override genes and cause 'pathological handedness'.

Why humans alone among animals are 90 per cent right-handed is a separate question, with the answer going back 2 million years. This is when human brains became asymmetric and the neural equipment for the fast, precise movements for speech and finger dexterity became localised in the left hemisphere. Why it is the left hemisphere is unclear.

Why aren't all animals ambidextrous? Most likely because it pays to specialise – if all practice is with just one hand, that hand will be more accomplished than a hand that only benefited from half the practice time.

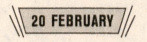

Why are blue whales so big?

BLUE WHALES NEED to be so big precisely because they eat tiny food, specifically krill, small crustaceans that feed on plankton.

Krill defend themselves against smaller predators by forming very dense shoals that confuse attackers. Blue whales get around this by swimming very fast at the shoal with their mouth wide open, often from below, and engulfing as much as they can. Their pleated throat can expand enormously to take in as much of a shoal as possible. The whales then strain the krill by forcing the water out through the baleen plates in their mouth. Because each attack uses up a lot of energy, this feeding strategy – called lunge feeding – only works if the animal can take in enough krill to more than repay the cost, so the whale needs to be huge to use it effectively.

Moreover, their huge size keeps them warm. A struggle for mammals that live in a cold ocean is to maintain their body temperature. Besides having a thick layer of blubber as insulation, another strategy is to be as large as possible, because this mini-mises their surface area-to-mass ratio. In other words, there is less surface area through which to lose heat per unit of body weight. Young whale calves don't have this benefit, which is one reason why many whale species migrate to warmer waters to give birth. The calves then have the chance to fatten up before they venture to colder waters.

The upper limit on body size is determined by the skeletal struc-ture needed to support the body. For land mammals this is a more serious limit because they don't have the buoyancy of the water to help support their weight. That is why land mammals can't grow as large as whales.

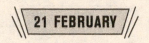

Can you make a tornado?

IN THE EARLY 1960s, a French scientist, J. Dessens of the Observatoire du Puy de Dôme, Clermont University, accidently discovered a way to make tornadoes artificially – and therefore a means of studying the conditions under which they arose.

On a plateau in the south of France the observatory built an apparatus which was originally intended for making artificial cumulus clouds. It was called the Meteotron and consisted of an array of 100 burners spaced over an area rather larger than a football field. Fuel was pumped into them and, together, they consumed about a ton of oil a minute, producing the very considerable power of some 700,000 kilowatts. In operation the device produced a thick column of black smoke that permitted observations of the resulting upward air currents.

During one experiment there appeared, besides the main mass of smoke, what seemed to be a black tube of whirling smoke. This tornado, about 30 or 40 feet across and up to 700 feet high, seemed to form about six minutes after the burners had been lit and subsequently moved away from the apparatus at the speed of the prevailing wind.

Later, the research team attempted to reproduce the tornado conditions with only half the burners operating, but under very unstable atmospheric conditions. After about a quarter of an hour's heating, they started a strong whirlwind in the centre of the apparatus about 130 feet across, with a bright tube a yard in diameter at its centre. It was so powerful that some of the burners were extinguished.

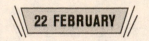

Can you see infinity?

PLACE TWO MIRRORS so that they are facing each other. Stand between them and look at your reflection in either mirror. Your reflection will seem to stretch into the distance in both mirrors, curving away to become smaller and smaller until you can no longer determine where it ends.

But because no mirror reflects 100 per cent of the light falling on it, sadly you can't see infinity. If it's a very good mirror and can reflect 99 per cent of the light, after about 70 reflections only 50 per cent of the light will remain, after 140 reflections only 25 per cent of the light is left, and so on until there is not enough light left to reflect between the two mirrors. Additionally, most mirrors reflect some colours of light much better than others, so the multiple reflections that you see not only get darker but they also become more colour-distorted as they recede.

Even with perfect reflection of all colours you could never see infinite reflections because positioning the mirrors so that they were perfectly parallel is practically impossible to achieve. This is why the images appear to curve away, until eventually the reflection is lost 'around the bend'.

Another obstacle is that your eyes are in the middle of your head, not at its edge. Therefore, at some point, the receding and hence apparently more distant mirror images would become smaller than, and hidden behind, the first reflected image of your head.

However, for those who prefer a theoretical mathematical explanation of whether infinity awaits us at the end of a reflection, there is hope. Assuming perfectly parallel mirrors, perfect reflection and a transparent viewer who was prepared to stand around for ever, infinity is just about achievable.

Take a deep breath – here's how it works. Light travels at a finite speed (c) – roughly 3×10^8 m per second. If your two mirrors were L metres apart and you stood between them for t seconds you would be able to see c multiplied by t divided by L reflections. For example, if your mirrors were 2 m apart and you stood between them for 1 minute, you should be able to see about 9 billion reflections. If you could stand around for ever, those reflections would presumably reach infinity . . .

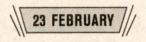

23 FEBRUARY

Why does your blood type matter?

IN THE ABO system, there are four blood types: A, B, AB and O. These designations refer to the types of sugars (A, B and O) found on the surface of red blood cells. Everyone has an O sugar, and those who have no other type are known as blood group O. The other group names arise from the fact that some people have A, B or both A and B sugars attached to the O sugar.

Your blood type is determined by the alleles (the name given to versions of the same gene) you inherit from your parents. The A and B alleles are dominant, while the O allele is recessive. If the gene from your mother is the recessive allele for producing O cell surface markers, and the gene from your father is for A cell surface markers, then your overall blood type is A. Why? Since everyone has O we only look at the second sugar present to determine blood type. If the gene from your mother was for B and the gene from your father was for A, then your blood type would be AB. If both parents donated A alleles, then you would be A. And if both gave you recessive alleles then your blood type would be O.

Cells use things such as proteins and sugars on their surface for many purposes. One is to enable the immune system to tell 'self' from 'non-self' and distinguish 'you' from every 'foreign' body that may invade. This has important consequences if you're ever in need of a blood transfusion. If you have type A blood, meaning that you have both A and O sugars on your red blood cells, you can accept both type A and O blood from a donor. If type B blood was given by mistake, your immune system would attack those blood cells, and the transfusion would kill you.

Type O blood is the universal donor, because nobody makes antibodies to this blood type – we all have the O sugar. Type AB blood is known as the universal acceptor because all blood sugars are recognised as self.

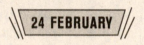

24 FEBRUARY

Why is fire red?

FIRE INVOLVES A chemical reaction between fuel and atmospheric oxygen. Once initiated it is self-sustaining, generates high temperatures and releases a combination of heat, light, noxious gases and particulate matter.

A flame is a region containing very hot atoms. At high enough temperatures all atoms will emit energy in the form of light as their electrons, which have been prompted to higher energy levels by absorbing heat energy, fall to lower energy states.

But fire is not always red. Because this light is emitted in discrete amounts according to the relationship $E = h\nu$ (where E = energy, h = Planck's constant and ν = frequency), flame colour is related to the magnitude of the quantity of energy which is transformed in to light.

This can most easily be seen with a Bunsen burner. A Bunsen burner that has a choked air supply burns cool, and the light emissions from carbon atoms are relatively low in energy and appear more red or orange. However, when the Bunsen is allowed air so that combustion is complete, the flame is hotter and the light emitted is of a higher energy and frequency and appears blue.

The luminescence of a flame is only half the story. The structure of the flame region is important to understand too. The flame area in a normal combustion environment, such as an open-air bonfire, is structured by convection currents which form as hotter, lighter air rises and allows cooler fresh air to replace it. It is this channelling effect and movement of air that shapes the dancing flames. It is interesting that in space, in zero gravity, the hotter and cooler air cannot move by convection, so flames take on weird shapes and may be stifled by their own combustion products.

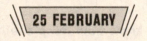

25 FEBRUARY

Why do dogs have black noses?

DOGS HAVE MOST likely developed black noses as a protection against sunburn. While the rest of the dog's body is protected by fur, light-coloured noses are exposed to the full force of the sun's rays. Pink-nosed dogs, hairless breeds and dogs with very thin hair on their ears need to be protected with sunscreen when they go outdoors, just as humans sometimes do, or they risk the same sorts of cancers and burns.

In addition, dog breeders have long singled out a black nose as the only acceptable colour for many breeds. Though this is based on nothing more than an aesthetic preference, it still serves as a selective influence for people breeding pedigree dogs. This adds a bit of human-directed evolution to what was already a natural tendency towards black noses.

While a majority of dogs have black noses, not all do. The noses of dogs such as vizslas and weimaraners match their coat colours – red and silver, respectively – and it is not unusual for puppies of any breed to start out with pink noses that then darken as the animal matures. Some dogs have pink noses all their life and are known as 'Dudley-nosed' because the genetic mutation was first noticed in bulldogs from Dudley, Worcestershire in the UK. Dudley nose is caused by a mutation in the *TYRP1* gene and only occurs when a dog is homozygous, or has two recessive alleles for the mutation. This means a pair of dogs could have black noses but be carriers of the mutation and so have Dudley-nosed puppies.

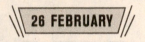

26 FEBRUARY

How do we know how deep the oceans are?

IN OUR CURRENT climate crisis, calculations of sea levels are a matter of great importance. But how are they measured when, even on a windless day, the ocean surface rises and falls?

Sea level is measured at the coast by tide gauges, and globally by satellite altimeters. The gauge can be an old-fashioned float type, or an acoustic sensor or radar. The rapid changes in height caused by waves can be smoothed out by mounting the tide gauge in a 'stilling well', essentially a vertical tube with an opening at the bottom that is about one tenth the diameter of the tube itself. The stilling well damps out the waves but allows tides and surges to be measured.

With satellite altimeters the picture is more complicated, as many adjustments have to be made to account for air pressure and water vapour content in the atmosphere, as well as scattering by waves and the effects of tides. The processing is very complex and even then the altimeters need to be calibrated with tide gauges.

For both tide gauges and satellite altimeters, the key to reaching millimetre accuracy is averaging. Satellites average over an area about 7 kilometres in diameter, and data from both satellites and tide gauges is averaged over time. In this way tides, waves, storms and even the seasonal cycle are removed from the data, allowing mean sea level to be determined very accurately.

Satellite altimeters have taken only a few years to map previously uncharted ocean bottom that would have taken 100 years to map using conventional techniques. The information this yields is priceless. For example, the altimeter maps have been used to identify an impact crater south of New Zealand 20 kilometres in diameter. Detection of this crater using ship-borne techniques would have been virtually impossible and too expensive.

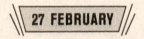

27 FEBRUARY

Why does brown bread toast quicker than white?

PLACE A SLICE of brown bread in the toaster. Check it every 15 seconds to see when it starts to burn (you'll need to introduce an objective measure of what you consider to be charred bread). Take a note of the time. Repeat with the white bread. Once you're done with your experiment, get out some butter or jam to ensure your efforts aren't wasted.

You'll notice that the brown or whole wheat bread toasts much faster than the white bread. During heating, a complex reaction occurs between the proteins and sugar contained in the bread. This is known as the Maillard reaction, and it produces the typical flavour we know from toasted bread as well as the colour formed as the bread toasts.

The Maillard reaction is a chemical one between an amino acid and a reducing sugar. It usually requires the addition of heat, as in the case of toasting bread. The reaction is widely used in the food industry for flavouring, producing different flavours and odours according to the type of amino acid involved in the reaction. Because brown bread and whole wheat bread contain more sugar and protein than white bread, they undergo more rapid Maillard browning than white bread.

One other factor may help to explain why white bread chars more slowly, and this is albedo – the proportion of incident light or radiation that is reflected by a surface. White bread reflects more radiation than brown bread, which is why it appears whiter. Because darker breads absorb more radiation in the form of heat from the toaster, they heat up more quickly and therefore burn quicker.

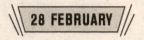

28 FEBRUARY

Why do we say 'um' or 'er'?

LINGUISTS KNOWN AS conversation analysts have observed that people vocalise in a conversation when they think it is their turn to talk, and there are several ways of negotiating the taking of those turns. One of them is the relinquishing of a turn by the current speaker and another speaker taking the floor. Therefore, silence is often construed as a signal that the current speaker is ready to give up his or her turn.

So, if we wish to continue our speaking turn, we often need to fill the silences with a sound to show that we intend to carry on speaking. In order to keep the floor while we think of what to say, we place dummy words in the empty spaces between our words, much as we might drape our coats on a seat at the cinema to prevent others from taking it.

As to why 'er' and 'um' are used instead of, say, 'choo', is not as easy to answer. 'Er' in British English is a transcription of the phonetic 'schwa' sound found in unstressed syllables of English words (such as the vowel sound in the first syllable of 'potato'). In traditional phonetics this was called the neutral sound because it is the vowel sound produced when the mouth is not in gear, that is, not tensed to say any of the other formed vowels such as 'i'. 'Um' is really only 'er' with a closed mouth.

These sounds are not universal. Many speakers of other languages hesitate in other ways. In Romance languages, for example, the pure sound of the vowel 'e' is often used. Mandarin Chinese speakers often say 'zhege zhege zhege' (this this this). Some young, hip foreigners learning Mandarin soon 'zhege zhege zhege' with the best of them.

English speakers might even argue that it's not 'um' or 'er' any more, it's 'I was like, omigod, that's like so totally not good . . .' Fluttering your hands in front of your face as if to cool yourself down at this point is, like, a totally optional extra.

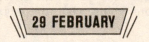

29 FEBRUARY

Is water the ideal liquid in which to swim?

THE SPEED OF a body in fluid is limited by the sum of three factors. The viscous drag is the friction of the fluid against the wetted surface. The form or pressure drag is the force created by the pressure difference between the front and the rear of the body. Finally, there is the wave-making drag, which is the energy wasted in making waves on the surface of the water.

Assuming there would be no ill effects to your health, there are a few options to achieve a higher speed for a given power. You could swim totally submerged in a liquid with a lower viscosity and lower density than water, such as acetone, methanol or ether. The lower viscosity would cause less friction and reduce pressure drag, which is proportional to the density of the fluid, cross-sectional area of the swimmer and square of their velocity. Swimming below the surface would totally eliminate the wave-making drag. This is what submarines do to achieve high speeds.

Another option is to swim on a liquid that has a much higher density than water but low viscosity. Mercury, which has a density 13.6 times that of water, would be ideal. A swimmer weighing 90 kilograms whose back has a surface area of 3,000 square centimetres could do a modified backstroke with their torso displacing less than one inch of the mercury. The swimmer could keep all of their limbs out of the liquid and use vigorous heel kicks into the mercury as an effective means of propelling themselves forward. Mercury does not wet skin and the sharp shape of its meniscus would further reduce the drag. With less than an inch of immersion, the swimmer's body would virtually 'hydrargyro-plane' across the smooth surface.

In a ceramic-fibre bodysuit a swimmer could do even better in a pool of molten gold, platinum or uranium, displacing barely half an inch of liquid – before frying when the thermal insulation of the suit failed.

MARCH

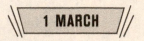

1 MARCH

Why do raspberries have little hairs?

THE HAIRS ON raspberries are the remains of the female parts of the raspberry flower, which have not fallen away. In the flower, the female hair-like styles are collected in the centre with the male anthers arranged around the edge. Each style, topped by a stigma, is connected to one ovary, forming a pistil.

After pollination, the petals, anthers and other parts of the flower wither away, and each ovary swells to produce a segment of the final fruit. Each of these segments is, botanically speaking, a drupelet, and a raspberry is a drupe (a collection of these segments) rather than a berry.

Raspberries are very easy to grow and very fruitful. If you grow your own you can see the full cycle throughout the same plant, from flower bud to ripe fruit.

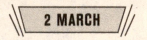

2 MARCH

How young can you die of old age?

LOOSELY, SOMEONE DIES of old age when ageing is responsible for the changes that have brought their body to terminal failure.

There's no escape from the processes of ageing; we lose the ability to repair and regenerate tissue with stem cells, and we accumulate DNA damage as well as old and dying cells.

So when does someone become old? It's a sliding scale, and experts joke that a general rule is that it's fifteen years from whatever your age happens to be right now.

But there are some more practical definitions too. One of these is multimorbidity. From around the age of seventy, we start to accumulate age-related chronic illnesses, like heart disease, arthritis, dementia and diabetes. When the Newcastle 85+ Study began in 2006, they found that, out of more than 1,000 eighty-five-year-olds, three-quarters had four or more medical conditions. When a person has multiple diseases, the cause of their death can be less clear-cut, and dying of old age starts to become a more useful term.

A defined, immediate cause of death is more likely to be listed on a person's death certificate – for example, cardiac arrest or pneumonia. But one definition of dying of old age could be that, if it hadn't been one cause today, then it might have been a different one tomorrow.

Everyone ages at a different rate, and it isn't possible to say exactly how young a person could die from old age. But statistical methods can help us work out how old you have to be before death from old age starts to become increasingly likely.

In this way, David Melzer at the University of Exeter, UK, and his colleagues can predict death due to ageing for particular demographic cohorts. There's a fair amount of variation, but their calculations suggest that the majority of deaths from old age will occur after the ages of eighty-one in women, and seventy-eight in men. So we can say that dying in your eighties can't yet be considered 'before your time' – although we suspect many octogenarians would disagree.

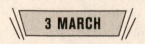

3 MARCH

Why do we need to wee more in cold weather?

IT'S FOR THE same reason that you feel thirsty after going for a swim. When you go out in the cold, or get into cold water, your body restricts the flow of blood to your peripheries and the skin by constricting peripheral blood vessels. This allows it to preserve heat and maintain core temperature.

One consequence of this is that the total volume of your blood vessels is reduced, which initially means your blood pressure increases. In response, your kidneys excrete more fluid and thus reduce blood volume, restoring your blood pressure to its previous level.

When you come into the warm again, the mechanism works in reverse and you have to either drink more or more water has to be taken up from your gut to compensate. In warm conditions we transpire to keep cool, losing water through our skin and mucous membranes, and in consequence excrete less urine.

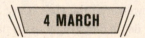

4 MARCH

Why do we get pins and needles?

PINS AND NEEDLES, properly called paraesthesia, can be caused by a number of things. It is commonly felt in the extremities and is usually caused by a lack of blood supply or by inadvertent pressure placed on a superficial nerve. For example, if you kneel or sit on your legs, the weight of your body tends to limit the blood supply to the lower limb and, as a result, the nerves become starved of blood and start to send unusual signals to the brain. This is perceived as a tingling sensation or pins and needles in the foot or lower leg. Once you move and change position, the nerve compression is released and the pins and needles gradually fade.

Paraesthesia may also be chronic. Poor circulation is common in older people. It may be caused by conditions such as atherosclerosis or peripheral vascular disease. Without a sufficient blood supply and hence nutrients, nerve cells cannot function normally. This is also why paraesthesia can be a symptom of malnutrition, as well as metabolic disorders such as diabetes and hypothyroidism.

Additionally, inflammation of tissue can irritate nerves running through it, causing paraesthesia. This is the case in conditions such as carpal tunnel syndrome and rheumatoid arthritis. Chronic paraesthesia can sometimes be symptomatic of neurological disorders such as motor neurone disease or multiple sclerosis.

5 MARCH

Why is frozen milk yellow?

THIS QUESTION WAS answered by the award-winning food writer, Harold McGee, who is the author of *On Food & Cooking* and *Nose Dive*.

The yellow colour of frozen milk comes from the vitamin riboflavin, which actually got its name from its colour – *flavus* is the Latin for yellow.

Riboflavin is dissolved in the watery portion of milk, which is also filled with minute particles of protein and droplets of butterfat. In fresh milk, all the suspended particles and droplets scatter any light that strikes them evenly, so that the milk appears opaque and white – milky, in other words.

However, as the milk freezes and most of its water crystallises into ice before other substances, the normally dilute riboflavin becomes concentrated in the remaining liquid water. This means these areas start to turn yellow and, as the clear water-ice crystals form, we are able to see it.

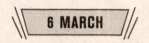

6 MARCH

How do lava lamps work?

THE ORIGINAL LAVA lamp was invented by Ronnie Rossi in the 1960s. It comprises a glass bottle sitting atop a light bulb that heats the contents of the bottle. Inside the bottle are water and coloured wax with a metal coil at the base. At room temperature wax is denser than water, but when it is heated it becomes less dense. So when the lamp is switched on and the bulb heats the bottom of the bottle, the wax becomes less dense and begins to rise through the water. When a globule rises, however, it cools as it moves away from the heat source, increasing in density and falling back down to the bottom of the bottle, where it starts to heat up again. The metal coil also helps the wax to recoalesce so that the process can restart.

This is a trendy effect and you can recreate it at home. All you'll need is a clean plastic 2-litre bottle, water, food colouring, vegetable oil and an effervescent antacid tablet, such as Alka-Seltzer.

Colour the water with the food colouring. About 10 drops will ensure the water will be quite dark – red or purple colouring gives a pleasing effect. Fill the bottle three-quarters full with vegetable oil, and top it up with the coloured water. Divide the antacid tablet into eight and drop one piece into the bottle.

Oil and water don't mix, so at first the oil – which is less dense than the water – will float above the coloured band of water at the bottom of the bottle. You'll also notice that the food colouring stays in the water and doesn't taint the oil. When you add the tablet it will fall through the oil to the water below and begin fizzing as it reacts with the water. Then you'll see globules of coloured water begin to rise up through the oil before reaching the surface and sinking again. When the reaction stops you can start it again by adding another piece of antacid tablet.

Antacids are made from sodium bicarbonate and citric acid. When these are placed in water they react vigorously. One of the key products of this reaction is carbon dioxide, plus water and various salts. So when the tablet reaches the water at the bottom of the bottle it reacts with the water, forming bubbles of carbon dioxide. These bubbles become attached to the oxygen and hydrogen molecules that make up water and, because the bubbles are much lighter than water or oil, they rise to the surface, dragging coloured liquid with them. Oil is viscous, so their journey to the surface is relatively slow, which means they rise with a pleasingly gentle action. When they reach the surface they burst, releasing the coloured water they dragged along for the ride and allowing it to sink once more to the bottom of the bottle. It is this constant rising and falling of coloured water that leads to the impressive lava lamp effect.

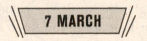

7 MARCH

Could we learn to photosynthesise?

If we could somehow splice the recipe that plants use for making chlorophyll into our genes, could we then satisfy some of our energy needs by photosynthesising?

First you would need to make the chlorophyll, then deploy it to harness energy from light for synthesising ATP and NADPH, and lastly use these metabolites to convert carbon dioxide into sugar. Surprisingly, we already have many of the genes needed for the first and last stages.

Plants use sixteen metabolic reactions to create chlorophyll, nine of which are shared with the pathway we use for making haem, a constituent of red blood cells. So inserting genes for seven extra steps could in principle allow us to make chlorophyll. Unfortunately, the chlorophyll would be toxic to humans, especially in sunlight, unless we also copied plants in making special proteins that envelop the chlorophyll molecules.

The last stage, known as the Calvin–Benson cycle, requires genes for eleven enzymes, nine of which we already have. The two missing enzymes are RuBisCO and phosphoribulokinase.

The middle stage is the trickiest because, although we already have the enzymes needed to make ATP and NADPH, we don't have a way of powering them by light energy. Plants have ingenious proteins and lipids, structurally organised within chloroplast membranes to allow this.

Before you try to engineer these botanical feats into your own chromosomes, be warned that you would be able to photosynthesise only minute amounts of carbohydrate compared with what you get from, say, a slice of bread. The problem is that your body would need a huge surface area relative to its volume in order to absorb useful amounts of light. Obviously, the chlorophyll would need to be in your skin; your liver, lungs and brain are not well placed to absorb light. The skin of an 85-kilogram adult provides a surface area of less than 2 square metres. In contrast, an 85-kilogram plant typically presents an area of over 200 square metres to the sun, thanks to its thin, expansive leaves. With an area to volume ratio like that, we would have great difficulty moving around – in fact, we'd be vegetative.

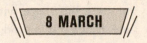
Why does halloumi squeak?

SQUEAKY HALLOUMI IS enough to make some people's toes curl, like fingernails dragged down a blackboard. This is because such sounds often warn of injury – a broken bone grating – or an unpleasant sensation, such as sand in your teeth, or stone abrading fingernails.

Probably long before our ancestors evolved into apes, they developed an inherited distaste for such noises and the associated sensations. It was likely an evolutionary adaptation to their way of life; those who did not respond to the signals tended to have shorter and less productive lifespans.

The squeak of halloumi is an example of the stick-slip phenomenon. The cheese is rubbery and as your teeth begin to squeeze it, the halloumi deforms with increasing resistance until it loses its grip and snaps back to something like its original shape. At the point where the slipping stops it regains its grip and the process repeats, commonly at a frequency near 1000 hertz, give or take an octave or two. The vibration produces a squeal of corresponding frequencies that may vary with the circumstances, such as whether the cheese has oil on it.

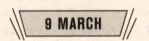

9 MARCH

How much vitamin D do you need?

VEGAN OR OMNIVORE, most of us depend on sunlight for meeting our vitamin D needs. A young fair-skinned adult should be able to synthesise enough vitamin D by exposing their face to the sun for about twenty to thirty minutes, but those with dark skin and the elderly will need more time. Older adults produce about 25 per cent of the levels young adults do for the same length of exposure. The amount of exposed skin matters too. Naturists should find it easier to meet their daily allowance.

Sunlight exposure is not, however, as straightforward as it seems, because only the narrow band of ultraviolet B (UVB) light (between 290 and 320 nanometres) is suitable for synthesis of vitamin D in the skin. The further you live from the equator, the harder it is to get an effective dose of sunlight, especially in winter. Winter levels of vitamin D in the blood are half those of the summer. If your shadow on the ground is longer than you are tall, then you cannot synthesise vitamin D (this is known as the 'shadow rule'). In Scotland for example, early morning sun in December is useless and there is no significant synthesis of vitamin D between October and March. In summer it would take someone in London around twice as long to make the same amount of vitamin D as a person sunning themselves in Barcelona, Spain.

It's important for people living in colder climes to supplement their diet with healthy sources of vitamin D, such as oily fish, eggs and liver. However, despite fortification of foods, including breakfast cereals and spreadable fats, the average dietary intake in the UK is poor – somewhere between 2 and 3.5 micrograms daily. Estimates of daily requirements vary, but UK guidelines suggest that adults who cannot get enough sunlight should take supplements of 10 micrograms (you should always check with your doctor before changing your diet or taking supplements).

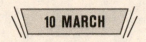

Why do we have fat?

MANY PHYSIOLOGICAL DETAILS of our adipose tissue, where fat in the body is stored, are poorly understood. However, adipose tissue varies greatly, so some people have more troublesome spare tyres than others.

One reason is that in our recent past there has been evolutionary selection for different patterns of accumulation of mature fat. In frigid climates bears and humans need layers of blubber, which in the tropics could be fatal. In contrast, aboriginal peoples in hot climates with intermittent food supplies tend to carry fat on the belly, buttocks or outer thighs, just as camels store fat on their backs.

There are functional differences between the fat of human babies and adults. Babies' 'brown fat', a specialised form of adipose tissue, counters hypothermia, and youngsters' puppy fat is readily mobilised for growth and activity. However, adults' love handles are for reproduction, hard times and famine, and to deplete strategically deposited stores too readily would be foolish.

So remember, fat is there for a purpose and as such should not be dismissed unappreciatively. It must manage fuel storage, conversion and mobilisation; it has elaborate endocrine functions, both metabolic and reproductive; and it differs drastically between populations as well as between people.

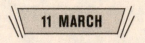

Why does lightning fork?

LIGHTNING USUALLY BRINGS the negative charge from a thunderstorm down to the ground. A negatively charged leader precedes the visible lightning, moving downwards below the clouds and through air containing pockets of positive charge. These are caused by point discharge ions released from the ground by the thunderstorm's high electric field.

The leader branches in its attempt to find the path of least resistance. When one of these branches gets close to the ground, the negative charges attract positive ions from pointed objects, such as grass and trees, to form a conducting path between cloud and ground. The negative charges then drain to ground starting from the bottom of the leader channel. This is the visible 'return stroke' whose luminosity travels upwards as the charges move down. Those branches of the leader that were not successful in reaching the ground become brighter when their charges drain into the main channel.

Photographs of lightning often overestimate the channel width because the film can be overexposed. Damaged objects that have been struck by lightning show channel diameters of between 2 and 100 millimetres.

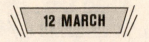

Why do we get morning breath?

DO YOU EVER wake up with less than pleasant smelling breath? Don't worry, you're not alone. And when you look at the cornucopia of life that lives inside your mouth, it's hardly surprising!

Our mouths are home to around 700 types of bacteria. As well as harmful organisms, which can cause tooth decay, gum disease and permanent bad breath, there are 'good' bacteria, which promote oral health by stopping the harmful ones proliferating.

During the night, your saliva flow slows and is less effective at washing out food particles and delivering oxygen to the bacterial flora. This stimulates the growth of anaerobic microbes, which are particularly smelly – hence bad breath in the morning. Bad breath is likely to be more pronounced if you have been breathing through your mouth, as this will dry out the saliva, further cutting the chances of a good wash-out.

This is similar to the smell that arises on an unwashed water bottle. When you drink directly from a bottle, you leave some of your oral bacteria and saliva on its neck. The saliva contains food debris and dead cells on which oral bacteria can thrive. If you don't wash the neck after you have drunk from the bottle, the bacteria left on the plastic will break down nutrients in the debris and release an unpleasant stale smell. The smell is always the same because your bacterial flora stays the same.

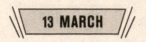

Why does electricity hum?

ELECTRICITY IS SUPPOSED to be the 'silent servant'. So why do electrical transformer stations and the like hum? To understand why, it is necessary to take a look at how a transformer works.

Transformers contain two coils of wire, the primary and the secondary coils, wound onto opposite sides of a ring made out of many thin sheets of iron or some other ferromagnetic material. An alternating current flowing through the primary coil generates an alternating magnetic field in the iron ring, which in turn creates a voltage in the secondary coil. The ratio of the primary voltage to the secondary voltage is equal to the ratio of the number of turns of wire in the primary coil to the number of turns in the secondary. This allows us to change the hundreds of thousands of volts running through overhead power lines to a voltage low enough to be safe to use in our homes.

The iron making up the ring that joins the primary and secondary coils is divided into microscopic domains. In each of these domains, the magnetic field points haphazardly in different directions, much like a classroom full of unruly pupils who are running all over the place. However, when the iron is placed in an external magnetic field, these domains tend to line up and add together, producing a strong magnetic field pointing in one direction, just as schoolchildren will snap to attention at a teacher's command (at least in the teacher's imagination . . .).

As the domains line up, the material very slightly changes its length to accommodate the rearrangement. This is magnetostriction. As the magnetic field through the iron alternates, the iron expands and contracts over and over again. These vibrations produce the sound waves that create the transformer's distinctive hum.

In the US, the mains voltage alternates 60 times every second (60 hertz), so the material expands and contracts 120 times per second, producing notes at 120 Hz and its harmonics. In Europe, where the mains supply is 50 Hz, the hum is nearer 100 Hz and its harmonics.

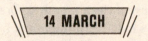

14 MARCH

Can you see your DNA?

YOU CAN! AND, as seems to be the case in so many *New Scientist* experiments, this involves a stiff drink. However, there's a perfectly good reason for breaking out the whisky – you can actually find out what makes you unique in the comfort of your own home . . .

Put a teaspoon of washing-up liquid diluted with three teaspoons of water into a clean glass. Fill another glass with water and stir in a teaspoon of salt. Swish the salty water around your mouth vigorously for 30 seconds or so then spit it into the diluted washing-up liquid. Stir this firmly for a few minutes, then very gently pour a couple of teaspoons of ice-cold strong alcohol down the side of the glass (a spirit of greater than 50 per cent alcohol by volume, such as a quality gin, strong vodka or rubbing alcohol, should do the trick). Use an eyedropper if you don't have a steady hand; tilting the glass also helps. This stage requires great concentration and is very important as you must have a clearly demarcated water/alcohol boundary. If you are careful, the spirit will form a separate layer on top of the salt/spit mix. Wait a few minutes and you'll see spindly, white, thread-like clumps starting to form in the alcohol. This is your DNA.

Swishing salty water around your mouth removes cells from the inner surface of your cheeks in the way you've seen in TV dramas when the police take a swab from a suspect's mouth for

DNA analysis. The detergent in the salt/spit mix breaks down the cells' membranes, releasing the DNA in the cell nuclei. Because DNA is soluble in water but not in alcohol it precipitates out in the white clumps you see floating on the surface. If you have a microscope you can investigate further, but if you just want to sit back and admire it, that stuff floating in the glass is what makes you who you are. But make sure you have a clean mouth before you start – if you've been scoffing a meat sandwich, it might not be your own DNA that ends up in the glass . . .

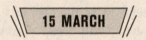

How do airplane toilets work?

TOILET WASTE IS moved by a vacuum through a waste line to a holding tank that is emptied on the ground at the airport after a flight. If there is not already sufficient vacuum, pressing the flush switch starts up a generator, which depressurises the waste line. This takes about a second to operate, during which time a rinse valve opens and then stays open for a further second. A small, measured amount of rinse water is used to clean the toilet bowl. Then, after this 2-second delay, a flush valve opens and stays open for a further 4 seconds to ensure the toilet bowl is empty. The change in pressure eventually moves the waste to the holding tank.

Toilets on an aircraft cannot be opened to the environment outside the plane for at least two reasons. First, flushing the toilet at altitude would cause explosive decompression of the cabin and second, if waste was scattered from the sky it would turn to ice and become a danger to people and structures on the ground. A frozen poo hurtling at speed towards you is probably not the last thing you want to see.

Conversely, waste water from the food and drink galleys and hand basins is dumped from drain masts that open to the outside. These are electrically heated but occasionally chunks of ice do fall from the sky following a malfunction.

16 MARCH

What is the storage capacity of the human brain?

IF THE HUMAN brain were like a computer, and each neuron held 1 bit of information then the brain could hold about 4 terabytes (4,000 gigabytes).

However, each neuron might hold more than 1 bit if we consider that information could be held at the level of the synapses through which one neuron connects to another. There are about 50,000 synapses per neuron. On this basis, the storage capacity could be 500 terabytes or more. But these are perhaps misleading answers because the human brain is not like a standard computer. First, it operates in parallel rather than serially. Second, it uses all sorts of data-compression routines. And third, it can create more storage capacity by generating new synapses and even new neurons.

The brain has many limitations, but storage capacity is not one. The problem is getting the stuff in and, even more problematic, getting the stuff out again. We can demonstrate that storage capacity is not the problem if we consider the technique experts use to remember the order of a shuffled pack of cards. This technique, called the 'method of loci', goes back to classical antiquity. It involves imagining a journey in which each card appears at a certain location.

If the first card, say, was an 8 of clubs, to memorise this you might imagine going out of your front door, the first step on the journey, and finding your path blocked by a person smashing an egg timer (which is shaped like an 8) to pieces with a mallet (a club). The next card is then placed on the next step of the journey with an equally vivid image.

What is striking about this technique is that the story you create to remember the order of the pack of cards contains much more information than the simple pack of cards you are trying to remember. The vivid images are necessary to get the information into our brain and to get it out again later.

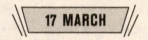

Why are Guinness bubbles white?

LOOKING AT A pint of Guinness there is no doubt that the liquid is black. Yet the bubbles that settle on top are white, despite being made of the same stuff. Why?

Unlike bath foam, which has many semi-coalesced bubbles, Guinness foam is made mainly of uniformly sized, spherical bubbles of about 0.1 to 0.2 millimetres in diameter, suspended in the good fluid itself. Just as a clear spherical marble, which has a higher refractive index than the surrounding air, can act as a strong magnifying glass, so spherical bubbles in beer diverge light because the air they contain has a lower refractive index than the surrounding fluid.

As a result, light entering the surface of the foam is rapidly scattered in different directions by multiple encounters with the bubbles. Reflections from the bubbles' surfaces also contribute to this scattering. Some of the light finds its way back to the surface and because all wavelengths are affected in the same way we see the foam as white. Light scattering from foam is akin to the scattering from water droplets that causes clouds to be white. This is called Mie scattering.

You can see this for yourself by pouring yourself a Guinness and placing some of the foam under a microscope. If you sit back and drain the glass, the drops of liquid at the bottom will have a light brown colour. Although Guinness appears black, it is not opaque. In the foam there is not so much liquid – most of the space is taken up by air. But because light is scattered from bubble to bubble the intervening brew does absorb some of it, providing a touch of colour.

Needless to say, to ensure reproducibility you may want to repeat the experiment several times – we recommend a Friday night.

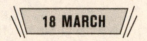

18 MARCH

Why do bees make honey?

BEES ARE HARDWIRED to forage for nectar, and if the nectar flow is good they will continue to fill every available space they can with stores of honey. Wild bee colonies with plenty of available space, such as in a large tree cavity or in a loft space, are capable of accumulating vast amounts of honey – enough to bring ceilings down.

Bees store honey for two reasons: first, to provide food to sustain them during flowerless periods, such as northern winters or long dry periods in tropical areas; and second, to produce a swarm – the only way colonies can reproduce.

If the workers decide it is a good time to swarm, they make a queen cell and have the current queen lay an egg in it. The grub is fed a special diet so that it develops into a queen capable of laying eggs. The old queen then leaves with half of the workers and drones. Before leaving, they ingest about half the stored honey. Beekeepers know just how much honey a swarm can remove – it is amazing that they can still take off.

To prevent their bees swarming, beekeepers have to open each hive every week and destroy any queen cells that have been built. If they fail to do this, then the hive will swarm. If humans stopped collecting honey, the colonies kept by commercial beekeepers would eventually disappear, and only wild colonies would remain. This would reduce the yield of many crops that use bees for pollination.

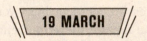

19 MARCH

How do Post-its stick to things?

POST-IT NOTES ARE a classic application of polymer chemistry. Funnily enough, the stickiness in question was discovered when Spencer Silver was trying to produce an incredibly strong adhesive in 1968. Instead, he produced an incredibly weak one, but most discoveries are accidents, so who's going to complain?

The glue used in Post-it notes is a pressure-sensitive adhesive. This means you only need to apply light pressure to stick the note to something. The bond between the note and the surface is formed through a fine balance between flow and resistance to it. The adhesive can flow just enough to fill tiny crevices on the surface, but will resist flow enough to remain there. This produces the bond between the note and the surface.

If we zoom in further, we can see that on a molecular level, the biggest contributors to the bond strength are van der Waals' forces. These are created when a molecule has more of its electrons on one side than the other, producing a dipole (like a tiny magnet). This induces the opposite dipole in another molecule nearby and the two stick to each other. Van der Waals' forces are normally weak, but they increase in strength as the size of the molecules grows.

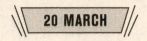

And how do they come off again?

WE KNOW HOW Post-Its stick to things – but how do they peel off again so harmlessly?

In practice, peeling leaves microscopic traces of Post-it glue on the target surface, but those are generally too small to cause visible damage. The secret is the molecular structure of the polymer, which forms a delicate network that links relatively large lumps together. Imagine it as being like a tennis net with sticky balls attached at intervals over most of its surface. The Post-it paper soaks up and holds powerfully onto one full surface of the adhesive mass (the tennis ball), but only one cheek of each sphere (or ball) sticks out, just touching the other surface.

This has two important effects: ensuring that cohesive forces greatly exceed adhesive ones, and that peeling the paper off the target pulls on only the next ball in line – each requiring only a small force at a concentrated point, almost like undoing a zip tooth by tooth. Strong contact adhesives are different in that they spread the peeling forces over as wide an area as possible. They make it hard to pull any point free without fighting all the neighbouring adhesive material at the same time.

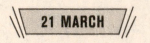

When are the equinoxes?

THE SPRING AND autumn equinoxes are defined as the point in time when the sun is overhead at midday local time on the equator (in astronomical terms, the time at which the sun crosses the celestial equator). On the equinoxes there is an equal length of day and night everywhere in the world. In the northern hemisphere the spring equinox usually falls on either 20 or 21 March and the autumn equinox on either 22 or 23 September (in the southern hemisphere the dates are reversed). This variation is simply because some years are leap years, so there is a shift in the calendar of a day or so relative to the seasons.

The equinoxes occur on exactly opposite sides of the Earth's orbit around the sun, but it is interesting that the dates on which they fall do not divide the year into two equal halves. Take the average dates of the equinoxes and the mean length of the year, and the autumn equinox falls 186 days after the spring equinox, whereas the spring equinox is only 179.25 days after the autumn equinox. This is because the Earth's orbit is elliptical and the Earth is closest to the sun in early January. In accordance with Kepler's second law, which states that a line joining a planet and the sun sweeps out equal areas in equal intervals of time, this is the part of the year when the angular velocity of the Earth in its orbit is greatest. As a result, the half of Earth's orbit from the autumn to the spring equinox takes less time to complete than the half between the spring and autumn equinox, when the Earth is further from the sun and moving more slowly. Consequently, spring and summer, during which there are more than 12 hours of daylight, last nearly seven days longer in the northern hemisphere than in the southern.

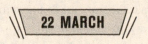

Could you survive on urine?

WE OFTEN HEAR about people surviving where water is scarce by drinking their own urine. While we hope none of our readers ever find themselves in a similar situation, what would happen if you needed a wee drink?

Well, if you must drink urine, your own is the safer bet. Somebody else's urine may not kill you, but it won't do you much good if it is carrying pathogens, pharmaceuticals or food allergens. Urine as produced by the kidneys is non-toxic and, urinary tract bacteria aside, your own will be free of anything alien.

The main solute in urine is urea, a harmless neutral end-product of protein metabolism and the ammonia this generates. The characteristic yellow colour is down to urobilin, the breakdown product of spent red blood cells. The urea will start oxidising back to ammonia on exposure to the air though, producing the smell characteristic of urinals.

One thing your body definitely won't want back are the salts that urine contains, which is why drinking urine is a bad survival tactic. It is not just the decreasing volume and increasing concentration with successive passes that is of concern. Osmotic effects mean that salts can only be excreted in solution – drinking these will make you more thirsty, hastening dehydration.

So while the nitrogen in urine makes it a great fertiliser and some insects will probe deposits for moisture and minerals, such drinking is best left to them. More effective in hot, dry situations is to soak an absorbent item of clothing with your urine and wear it as a hat, so providing evaporative cooling and some shade.

If the quantity of urine you intend to consume is particularly large and dilute because you gulped the last of your water too quickly, then drinking a first pass from your kidneys might conceivably help. The most efficient way to consume it is to drink about a mugful each time you start to feel thirsty. Then it won't enter the blood faster than the tissues can extract it.

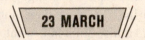

Why does spaghetti always snap into three pieces?

THIS IS, INDEED, a strange phenomenon. Surely holding a strand of dried spaghetti at both ends and bending it until it breaks should produce just two pieces, but it hardly ever does – usually three or even more pieces are the result. This conundrum first appeared in *New Scientist* in 1995 and was repeated in 1998 and 2006. It's a problem that has taxed greater minds than ours, including that of the Nobel prize-winning physicist Richard Feynman.

However, it was Basile Audoly and Sebastien Neukirch who verified what was going on in their paper 'Fragmentation of rods by cascading cracks: Why spaghetti does not break in half', published in *Physical Review Letters* (vol. 95, p. 95505).

Audoly and Neukirch broke strands of spaghetti of varying thicknesses and lengths by clamping one end and bending them from the other. They found that the unexpected three-part breakage occurs because of what are known as flexural waves. When the curvature of the spaghetti reaches a critical point, the first break appears. The shock of this causes a flexural wave to ripple down each of the two resulting lengths of pasta at high speed and amplitude.

The two halves formed by the initial break do not have time to relax and straighten before being hit by the flexural wave, which causes them to curve even further and suffer more breaks, leading to a cascade of cracks in the pasta. Often more than three pieces are created when this happens. While spaghetti snapping is in itself a rather humdrum if fun pastime, Audoly and Neukirch's work also provides important information about failures in other elongated, brittle structures, including human bones and bridge spans.

In *No Ordinary Genius*, the illustrated biography of Richard Feynman published in 1994, Danny Hills describes his and Feynman's numerous experiments with spaghetti that left them with 'broken spaghetti all over the kitchen and no real good theory about why it breaks into three.' This seems to have been a common occurrence – apparently visitors to Feynman's home were often presented with sticks of spaghetti and asked to help solve the problem.

There is, therefore, a delicious irony in the fact that while this puzzle drove Feynman, a physics Nobel prize winner, to distraction, those who discovered the reason why it happened – Audoly and Neukirch – were awarded the antithesis of Nobel fame, the Ig Nobel Prize for Physics in 2006, 41 years after Feynman won his Nobel prize. Nobel prizes are awarded for supreme achievements in scientists' chosen fields while Ig Nobels, from the opposite end of the research spectrum, are awarded for success in the areas of improbable research, humour and, quite often, silliness.

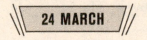

24 MARCH

Why don't we fall out of bed more?

GIVEN THAT THE average person twists and turns up to 100 times during a night's sleep, it is amazing that we don't fall out of bed more often.

At the University of Edinburgh, Geoffrey Walsh and *New Scientist* reader John Forrester investigated the reasons why adults do not usually fall out of bed while asleep. Volunteers slept on a very wide mattress in a warm room, with no coverlet so that they would not be able to detect in their sleep where they were in the bed. Their head position was noted from a choice of four positions: nose to left, nose up, nose to right, or nose down. The apparatus was unsophisticated, and comprised a rugby scrum cap onto which Forrester stitched a circle of plastic tubing complete with a short piece of glass tubing. The tubing contained some mercury, and he thrust some needles through the tubing wall at suitable points and attached a dry battery so that small voltages were generated according to the head position. These were recorded all night on an electroencephalographic recorder.

Periods of sleep and waking were recorded by arranging for a small sound to be made at around 10-minute intervals. If the volunteer was awake and heard it, they pressed a bell push attached to their clothing. This allowed the researchers to discount movements made during this time. During sleep, of course, no response was recorded.

Participants turned at irregular intervals throughout their sleep, for example, nose to left, nose up, nose to right, then back again. But they never turned nose down. As a result, they did not roll over and over so that they would fall out of bed. Instead, they remained in roughly the same position all night. The researchers concluded that quite early in life we learn that it is difficult to breathe if we turn nose down, and we avoid it even when asleep. As a result we won't roll out of bed.

If you find not falling out of bed in your sleep an impressive skill, spare a thought for sailors. In some ships they still sleep in hammocks; and the naval version, called a mick, is slung tight and level. Though it does seem to reduce sensitivity to the ship's motion, any sleeper who cannot lie flat and still in a mick is a hostage to fortune. It dumps you instantly if you so much as breathe asymmetrically, and yet thousands of sailors have slept soundly in them for centuries.

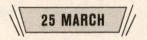
Why do zebras have stripes?

According to Rudyard Kipling's *Just So Stories*, the zebra got its stripes by standing half in the shade and half out, 'with the slippery-slidy shadows of the trees' falling on its body. While there have been many theories proposed by scientists over the years, the front-running hypothesis is that stripes serve mainly to repel insects.

Susanne Åkesson and her colleagues at Lund University in Sweden suggest that horseflies are attracted to linearly polarised light. Uniformly coloured animals reflect linearly polarised light, so that makes them a target. Zebra stripes disrupt the polarisation of reflected light, making them more difficult for insects to home in on. Female horseflies need to suck blood in order for their eggs to develop, and biting insects can transmit fatal diseases, so being able to evade them is an advantage.

The researchers tested their hypothesis by standing variously striped and coloured models of zebras, horses and asses in a field. They covered the models in glue, then counted the number of insects that became stuck. The zebras attracted the fewest flies.

Other researchers have lent support to this idea. One theory is that the prevalence of patterning on the hides of such animals increases in proportion to the population of biting insects. It has also been suggested that animals that evolve in areas where tsetse flies are present become striped.

No less a biologist than Alfred Russel Wallace suggested that the pattern also provides camouflage. Zebras are most vulnerable when drinking at watering holes, but if they do this at twilight, the stripes merge to a less conspicuous grey. Another take on camouflage is that lions are partly colour-blind, so do not perceive the contrast between zebras and the savannah in the way we do. When zebras are running as part of a herd, the stripes make it harder for lions to pick out individuals.

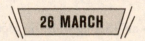

26 MARCH

Why are dock leaves good at relieving nettle stings?

IF YOU'VE EVER grazed a bare leg or arm on a nettle bush while out in nature, you'll know how much it stings. You might be told to find a dock leaf and rub it on – but why?

Being stung by a nettle is painful because the sting contains an acid. Rubbing the sting with a dock leaf can relieve the pain because dock leaves contain an alkali that will neutralise the acid and therefore reduce the sting. Bees and ants also have acidic stings, so dock leaves should help, but other alkalis, such as soap or bicarbonate of soda, are usually better.

However, a dock leaf is useless against wasp stings, which contain an alkali. This is unfortunate because wasps are nasty little critters whose sole aim in life is to ruin picnics and barbecues. If you want to neutralise a wasp sting you should use an acid such as vinegar. The only problem is you'll smell of pickles for the rest of the day.

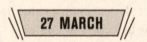

27 MARCH

How does a bike stay upright?

IN 2011, AN international team of bi-pedal enthusiasts dropped the bombshell that, despite 150 years of analysis, no one knows how a bicycle stays upright. What cyclists have been doing for years was a feat inexplicable by science.

Well, sort of. Andy Ruina, an engineer at Cornell University in Ithaca, New York, admitted that scientists don't know what the simple, necessary or sufficient conditions for a bicycle to be stable are. Instead, they've relied on trial-and-error engineering to construct stable bikes that aren't prone to toppling while in motion. Explaining how they work mathematically requires around twenty-five variables, such as the angle of the front forks relative to the road, weight distribution and wheel size.

Before 2011, researchers had reduced this profusion to two things. One was the size of the 'trail', the distance between where the front wheel touches the road and where a straight line through the forks would meet the ground. The other was the gyroscopic restoring force that acts on a spinning wheel to keep it upright.

Ruina and his colleagues, including Arend Schwab of the Delft University of Technology in the Netherlands and Jim Papadopoulos of the University of Wisconsin-Stout at Menomonie, not only re-visited this mathematics, but also skewed the trail and gyroscopic forces in prototype bikes to make them technically un-rideable. To everyone's surprise, the bikes were still stable.

'The real understanding of bikes requires a mix of mathematics, plus some kind of brain science,' says Papadopoulos. Human riders act in extremely complex yet intuitive ways to keep a bike balanced and on track. At very low speeds, for example, we recognise that the handlebars become useless for steering, and instead direct the bike by wobbling our knees.

Why? 'We don't know,' says Schwab. A bike-based mystery that could be around long after we've worked out the origins of the universe.

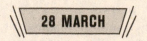

Why don't adults enjoy dizziness like children do?

CHILDREN OBVIOUSLY ENJOY the feeling of dizziness – just look at how roundabouts in parks and playgrounds are packed with youngsters. They need that stimulation to develop a healthy balance system, which is necessary to crawl, walk and keep their bodies upright, even on a rocking boat.

Our balance system is controlled by three senses cooperating in complex harmony. The vestibular system in our inner ear informs us about the position of our head; our eyes tell us how our body is located in relation to the external world; and proprioceptors – receptors in muscles and joints – help us to figure out how our body is positioned in space, which is particularly helpful if we cannot see. These elements mature at different rates.

The vestibular system is fully operational by the time a child has reached 6 months of age; proprioceptors need three or four years more. The development of the visual element is complete by around 16 years of age.

The sensation of dizziness and nausea following a spinning movement is similar to motion sickness – a result of the conflicting information our brain receives from the three elements mentioned above.

When our body is rotating at speed our vestibular system and proprioceptors can feel it, but our eyes can't locate the horizon. Our brain is desperately trying to resolve this conflict and, because humans are primarily visual, it assumes that the other senses are hallucinating, probably because of intoxication. So the brain tries to get rid of the assumed poison by provoking vomiting.

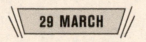
How high up would you need to be to see the curvature of the Earth?

As THE RADIUS of the Earth is 6,373 kilometres, a little trigonometry tells us that if you are at the top of a tower of height h metres, the horizon will be at a distance of approximately $(2 \times 6373 \times h)^{1/2}$ kilometres.

It's often said that the curvature of the Earth can be seen from the top of Blackpool Tower in the UK, but this might be a little tricky. For a tower 150 metres high, the horizon will be 44 kilometres away and displaced downwards from a true horizontal line by about 0.39 degrees. If you hold a 1-metre stick horizontally 1 metre in front of you, seemingly touching the horizon at the midpoint of the stick, the ends will appear to be 0.8 millimetres above the horizon. That's pretty hard to see with the naked eye.

The short answer to this question is that the curvature is not obviously visible from anywhere on the Earth's surface. Pilots of Lockheed U-2 and SR-71 Blackbird aircraft suggest that the Earth's curvature only becomes clear at an altitude of about 18 kilometres. Indeed, it has been photographed from Concorde cruising at this altitude. The curvature can be inferred at sea level, though. For example, ships disappear over the horizon from the bottom upwards, as if sinking into the sea.

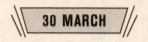

What causes dew?

SOME MORNINGS LAWNS can be covered with glistening drops of water; up close, these drops of water seem to occupy a precarious position at the very tips of grass blades. What causes this plant 'sweat'?

The process is called guttation. On the surface of leaves there are stomata or pores through which water is lost by transpiration. Plant roots take up inorganic ions from the soil and transfer them to the xylem from which they can't leak back. Water is drawn in by osmosis, which creates positive pressure in the xylem. Because of this pressure, xylem sap leaks from pores (the hydathodes) at the tips of grass leaves (or directly from the cut or cropped ends of the leaves). When the drops are large enough they fall and new ones form. Guttation usually happens at night because, by day, water loss from the leaves is normally sufficient to maintain a negative pressure in the xylem.

Guttation has been observed in more than 330 genera of 115 plant families, including potatoes, tomatoes and strawberries on their leaf margins. Among plants in temperate regions, species of Impatiens and grasses, including cereals, readily exhibit guttation. In the tropical plant *Colocasia antiquorum* 200 millilitres of water may be exuded by a single leaf in just one day.

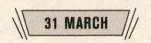

Are all calories equal?

IF YOU'VE EVER thought about losing weight, the idea of 'good' calories and 'bad' calories might sound very appetising. Is it too good to be true?

Yes and no. The calories are the same. If you restrict the number of calories in your diet, you will not show any difference in weight loss whether you restrict calories from fat or calories from carbohydrates. Once your body has converted the consumable compounds to a common compound (acetate, say) then there is no difference between the foods that were converted, whether protein, fat or carbohydrate.

However, this does not mean you will experience the same success with a diet that restricts fat as with one that restricts carbohydrates. We react differently to different foods, and fats help us feel satiated. A balanced, calorie-restricted diet is the best way to diet. Too little fat and you will often feel hungry; too little carbohydrate and you will feel tired and weak. Also, keep in mind that high-fat, high-protein diets place you at a greater risk of heart disease.

APRIL

Is there a scientific way to pick lotto numbers?

To ENTER THE national lotteries of many countries, you select six numbers. Many of us feel that we must choose at random, yet the odds would be the same if we always chose the numbers 1 to 6. Why do we feel this way? And is there a scientific way to pick lotto numbers?

Our brains cope with a large number of individuals by stripping them of their individuality, so we regard all sets of numbers which lack an obvious pattern as being essentially similar, and do not differentiate between them. Any set that does possess an obvious pattern feels like an unlikely candidate for winning because we perceive it as distinctive, special and – by perverse logic – unlikely to be drawn.

Despite the fact that 1, 2, 3, 4, 5, 6 is no less likely to win than any other set of numbers, if you're really hoping to win big money, this might not be the best set to choose. Some estimates show that if all six were ever to come up together, the jackpot would have to be shared between as many as 10,000 people because these people have chosen these numbers after also seeing through the meaninglessness of the patterns we like to impose on random numbers. What they have not realised is that there are a large number of like-minded people with whom they will have to share their winnings.

In this respect, the instinct to choose numbers 'at random' may yet have some rational justification. Probably the best tactic is to choose numbers that other people shun – there are a number of websites that offer advice on which numbers are more or less likely to appear and in what months or weeks of the year. You can choose to take the advice or ignore it, bearing in mind the draw is completely random.

Do animals recognise themselves?

THE QUESTION OF whether an animal, when looking in a mirror, recognises itself is still not a completely solved issue.

Experiments in primate self-recognition were done in the 1970s by Gordon Gallup. He anaesthetised chimpanzees, then marked their faces with blobs of non-toxic paint. The chimps were given mirrors to look in when they woke up. They saw the blob in the mirror, touched it and cleaned it off. Following Gallup's experiments, many researchers found behavioural evidence for self-recognition in apes (chimps, bonobos, gorillas and orangutans) and maybe in dolphins, orcas and even magpies.

Besides the experiments that involved marking faces, others were devised in which the animal was urged to do something with its hand, which was hidden behind a screen, while being shown a television picture of its hand in reverse. Chimps learned to manage the task by looking at the screen and moving their hand in a trial-and-error strategy guided by visual feedback. Other tasks investigated whether the animal recognised itself in a videotape image, a photograph, or even its own shadow.

Of course, all these experiments are behavioural tasks which cannot access the mind states of these animals directly and prove if they really do recognise themselves. Some researchers are therefore still sceptical. Macaques and other less intelligent primates have not yet displayed the self-recognition abilities of apes and humans. And apes that grow up with very limited social contact seem to fail the task too. Furthermore, developmental psychologists found evidence that self-recognition correlates with empathy. It is possible that both abilities emerge together. Incidentally, human babies normally learn to recognise themselves in a mirror at between 12 and 20 months.

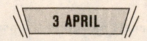

3 APRIL

Could we ever run out of words?

THE EXISTENCE OF polyglots suggests that the average person is far from running out of storage space in the brain. And while there are major structural constraints on word numbers, they too leave lots of room for expansion.

First, there is some effect from the number of distinct sounds (phonemes) in a language. Languages with fewer phonemes have some tendency to have longer words. English has about 40 phonemes and many short words, while Hawaiian, with only 13 phonemes, has many words of three or four syllables. Any language can increase the potential size of its vocabulary by using longer words.

Much more limiting than the number of phonemes are a language's phonotactic patterns – the constraints on possible sequences of phonemes. In English, words can begin with sequences like 'sp' or 'st', but in Spanish they cannot, hence the vowel at the beginning of Español. In Greek, words can begin with sequences like 'pn' or 'ps', but in English this is not permissible, so we pronounce Greek loan words like pneumatic and psychology without the 'p'.

Even so, we are a long way from using up all the permitted shapes: 'snizz' or 'whask', for example. Assuming that we strictly limit ourselves to only consonant–vowel–consonant forms and do not include such extras as tones and stress, the following provides a generous lower boundary to the number of words available in English.

There are more than 50 possible initial consonants (including combinations such as 'tr' and 'sk'). There are more than 10 distinct vowel sounds. Consider: rad, raid, red, rid, ride, rude, rod, reed, road, and add in non-words such as roid (void) and rould (should).

There are more than 40 terminal consonants (including 'rt' and 'lk'). This means that there are in excess of 20,000 (50 × 10 × 40) single syllables. If we limit ourselves to using only two syllables per word we would still have more than 400 million words to play with.

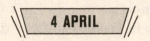

4 APRIL

Why do birds and bats perch differently?

BIRDS PERCH STANDING up, bats upside down – or do they?

There is a misconception that bats can't take off from an upright position, owing to their small leg bones and muscles which have been reduced to make flight more efficient. However, flapping their tail membrane allows bats to launch upwards. That said, such an ungainly take-off would make them vulnerable to predation if they nested on the ground, which makes dropping into flight from a perch a better strategy.

It's not just risk of predation that explains the difference though. Bats also have superior aerobatic ability. These animals can invert and come to a virtual standstill in flight which allows them to grasp a suitable roosting position from underneath. This means bats can monopolise the ceilings of caves and other inaccessible roosting sites. They are more manoeuvrable because their wings are larger relative to their body mass than is the case for birds. In addition, though wings have evolved from arms in both cases, the bone is limited to the front edge of a bird's wing, whereas in bats the fingers extend across and to the rear edge of the wing, giving bats finer control of the wing surface.

Conversely, birds' lack of manoeuvrability might limit them to more accessible roosts. And this may help explain why birds don't sleep as we understand the term – they rest each side of the brain in turn so that they stay alert to predators, and possibly to avoid falling off their perches.

However, there are a few birds including the vernal hanging parrot (*Loriculus vernalis*) that roost upside down in trees. The tendons in bats' and birds' feet are arranged to close the claws and lock the feet to the perch when the creature is relaxed, minimising energy expenditure. If a bat dies in its sleep, it doesn't automatically fall to the ground, and needs to be knocked off its perch.

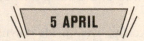

5 APRIL

Why does foil touching a filling hurt?

IF YOU HAVE looked after your teeth well, you can smugly sit this one out. Those with amalgam fillings, however, will quickly become enlightened.

Generate a decent amount of saliva in your mouth, pop some aluminium foil in and position it over a filling in your teeth, preferably a molar so you can bite on the foil. You may jump! You'll feel anything from a slight tingling in the filled tooth to actual pain as the foil touches it.

This is because when two dissimilar metals are separated by a conducting liquid, a current will flow between them, and this can stimulate nerves. In this case the two dissimilar metals are the amalgam in your tooth and the aluminium foil. A thin film of saliva separates the foil from the filling and, because it is a reasonable electrolyte containing various salts, it acts as the conducting liquid allowing a current to flow between tooth and filling. Because the filling is close to the dental nerve, the current will stimulate it, causing pain.

Luigi Galvani first discovered the dissimilar metal/electrolyte effect in 1762, when he carried out experiments using the nerves in frogs' legs. When probes of differing metallic composition were applied to the frogs, their legs twitched. Using tooth amalgam and spit is a more humane way of carrying out this experiment, although the owner of the filling might not agree.

6 APRIL

Could all the water in the universe put out the sun?

FIRE IS A chemical reaction that needs three key ingredients to keep going: the first is heat (such as from the match used to light a candle), the second is fuel (the candle wax), and the third is oxygen (there's plenty of that in the air). Remove any one of those three things and the flame will die.

Dousing fire with water is effective because water is very good at removing heat and cutting off oxygen. However, our sun is not actually made of fire like a giant candle. It is a giant ball of plasma. Rather than combustion, the sun runs on a process called nuclear fusion, where the heat and pressure in the sun's core are so vast that small nuclei such as hydrogen are forced to fuse into larger nuclei such as helium, generating blistering amounts of energy and keeping us at a comfortable temperature here on Earth.

Enveloping the sun in a thick blanket of water wouldn't be much help if you wanted to snuff it out. Although the water would instantaneously remove some heat it would also increase the sun's mass, and therefore the pressure inside it, increasing the rate of nuclear fusion. What's more, the water molecules (consisting of hydrogen and oxygen) might get hot enough to be ripped apart into their constituent nuclei, providing more fusion fuel. So the sun would actually burn more fiercely and rapidly than before.

But what if you dumped all the water in the entire universe over the sun? Now you're talking. Strictly speaking, you'd be dumping ice rather than water, because space is so cold that almost all water exists in its solid form. You would theoretically be able to add so much to the sun's mass in the form of ice that it would use all its fuel very quickly. Then it would explode cataclysmically as a supernova, destroying Earth and leaving behind an extremely dense neutron star, or even a black hole. I guess you could count that as extinguishing the sun. In summary, getting the sun wet is likely to seriously mess up our solar system.

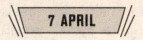

How do Aero chocolate bars get their signature bubbles?

IN ONE AERO (a famous brand of chocolate bar), there are approximately 2,200 bubbles in a chocolate matrix. The way in which the unique Aero bubbles are added is a top-secret process closely guarded by Nestlé Rowntree. But while the secret details may be absent, the broad answer can be found in Rowntree's British patent GB 459583 from 1935.

The chocolate is heated until it is in a fluid or semi-fluid state, then it is aerated, for example using a whisk, to produce many tiny air bubbles distributed throughout the chocolate. This is poured into moulds and the air pressure greatly reduced as the chocolate is cooled. The reduced air pressure causes the tiny bubbles to grow and gives the finished chocolate its frozen bubbled appearance. The solid chocolate coating on the surrounding surfaces of the bar is placed into the mould before the aerated fluid chocolate is poured in.

The patent gives no clues as to how the bubbles are prevented from rising to the surface during manufacture, but this may be due to the high viscosity of the semi-fluid chocolate and the rapid rate of cooling.

Patents provide a great source of technical information. It has been suggested that 80 per cent of technical disclosures appear in patents and nowhere else. You can view and print GB 459583 using the service on the Patent Office website, www.patent.gov.uk. The service provides an interface to British and European patent offices for you to search their databases. Two *New Scientist* readers picked up on another patent, GB 459582, which was filed by Rowntree on the same day as the one mentioned above and contains the 'Aero' concept. The chocolate makers clearly knew what they were about. Khachikian points out that eight days before lodging the patent, the Aero name was trademarked. Although British patents expire 20 years after they are filed, the trademark on the name Aero is still in force.

8 APRIL

What would happen if you lived in zero G?

LIVING ON A space station would be no picnic for your body. It's true you would be moving faster and so benefit from time dilation, according to Einstein. But a 6-month stint would only leave you

7 milliseconds in credit. Balance that against all the exercise you would need to do to fight muscle atrophy and the unavoidable loss of bone mass, both signs of ageing. If you were to venture through the Van Allen belts or orbit beyond them, you would receive an increased dose of ionising radiation from cosmic rays and the solar wind.

It's an appealing thought that, without the effects of gravity, we might evade wrinkles by living in zero G. However, it's likelier that you would end up looking like a featureless bag of fat and fluid with porous bones and spindly attachments where your limbs and head used to be because human bodies would be free to expand in any direction without the pull of gravity.

Overexposure to the sun, especially at lower latitudes on Earth, can cause quite young people to look like leather saddlebags with eyes, but you don't have to go into space to avoid that look. Gravity on Earth, however, has little to do with making us look old; our bits sag once they begin to lose fibre, tone, or their internal padding after first stretching to accommodate fat, growth, or muscle. Gravity is just one thing that shapes our structure and can't be blamed for the consequences of the other abuses our bodies suffer over our lifetimes, any more than you can blame food for obesity.

9 APRIL

How can an unopened jar of runny, clear honey turn into a hardened block of sugar?

BEE-KEEPERS ARGUE ABOUT this, as honeys from different sources behave differently. Honey is a supersaturated solution of various proportions of sugars (mainly glucose and fructose), and is full of insect scales, pollen grains and organic molecules that encourage or interfere with crystallisation. Glucose crystallises readily, while fructose stubbornly stays in solution. Honeys like aloe honey, which is rich in glucose and nucleating particles, go grainy, while some kinds of eucalyptus honey stay sweet and liquid for years.

Unpredictably delayed crystallisation means a nucleation centre has formed by microbes, local drying, oxidation or other chemical reactions. Crystallisation can also be purely spontaneous, starting whenever enough molecules meet and form a seed crystal. Some sugars do this easily, others very rarely.

By seeding honey with crystals, or violently stirring air into it, you can force crystallisation. Products made this way are sold as 'creamed' honey. The syrup between the sludge crystals is runnier and less sweet than the original honey, because its sugar is locked into crystals. Gently warm some creamed honey in a microwave until it dissolves, compare the taste of the syrup with the sludge – you will be astonished.

10 APRIL

Why do we like shiny things?

ARGUMENTS ON THIS point are necessarily speculative, but it's worth noting that we associate shininess, cleanliness and crisp outlines with objectively favourable attributes. In assessing a mate, a companion or a rival, we spontaneously see bright eyes and teeth, glowing skin and glossy hair as signs of health and quality. As children, we like things that stimulate our nervous systems with clear, vivid colours, contrasts and light.

Art may be seen as a form of play behaviour, in that it relies on elements that matter to our mental and physical development. As adults, our senses and creativity put a premium on media and themes that stimulate our innate mental systems in important ways.

Shiny things present intense, characteristic stimuli, and are used in social signals and communication, even in creatures that do not see art in our terms. Such dramatic signals may be based on anatomy or physiology, such as peacock tails or the belling of a stag, or may be collected and arranged as adornments, like bowerbird displays and human medals or finery. Much as we enjoy speech, we enjoy communication by vivid stimuli in a broader range of contexts.

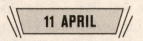

Why do muscles sometimes feel stiffer days after exercise?

IN ORDER TO improve performance when we exercise we need to progressively challenge our muscles with the amount of work we expect them to perform. This progressive overloading (usually achieved by increasing the resistance they experience such as lifting heavier weights or by running extra distance on successive days) causes tears in the muscle's microfibres. And, in a gradual overload/repair cycle, we experience moderate soreness up to a day later.

Delayed onset muscle soreness (DOMS) is caused when the expected load dramatically increases, causing a greater number of tears (rather than an increase in the magnitude of each tear). In this situation it takes longer for scar tissue to form because it grows in perpendicular fashion across the repair sites. Once the new tissue is in place, we experience the soreness that comes with DOMS as we reactivate and stretch this new, less pliable muscle, until its strength and flexibility are restored.

The sensation of discomfort usually develops approximately 24 hours after exercise, peaks at about two days and then gradually subsides. During the 24 to 48 hours post-exercise period, muscle swelling and stiffness usually result in a reduced range of motion and also muscle weakness.

12 APRIL

What should you do if you're trapped in quicksand?

THICK SLURRIES OF beach sand are often dilatant. Dilatant liquids are utterly perplexing because the harder you stir them, the more solid they become. Stop stirring and you have a liquid again. These positively mind-bending materials lend themselves to that old favourite of film directors: the victim trapped in quicksand. As the person struggles the sand takes a firmer grip. But using just cornflour and water, you'll now know exactly how to defy Hollywood legend.

Mix 300g corn flour and 250ml water in a medium-sized metal bowl until it becomes too hard to move the spoon. Stop stirring and tip the bowl. You'll notice the mixture becomes a liquid again and flows. Stir again vigorously and it will thicken once more. Try dipping a finger or the spoon into the mixture very slowly – if you are extremely gentle it will remain liquid, but pull out your finger or the spoon in a hurry and it will solidify. Drag your fingers quickly through the mixture and you'll be able to lift out a putty-like ball that you can work in your hands. But if you open your fingers and stop roiling it, it will quickly become a liquid again, so keep your hands over the bowl. Now try striking it with a hammer (this explains why you need a metal bowl). If you time the blow just right, the mixture will shatter. Best of all, the broken pieces will re-liquefy and pool together like the shape-shifting creatures in *Terminator 2*. You can even throw the material against an outside wall and watch it shatter (this bit can get quite messy).

The mixture you have created is a dilatant material, or a

shear-thickening fluid (STF). In these materials viscosity increases as the force, or shear, on them is increased. So, as you have already discovered, the more pressure you apply, the more resistant to deformation they become. This is because the application of force to the material causes it to adopt a more ordered structure. Under normal conditions the particulates in the liquid are only loosely arranged, but the shock of any impact or pressure changes their alignment, locking them into place. When the stress dissipates the material relaxes again.

Research into STFs has led to the development of 'smart materials', ones that respond to changes in their surrounding environment. For example, military researchers are attempting to treat fabrics with STFs so that when a bullet strikes an STF-treated uniform, the uniform becomes rigid at the point of impact and the bullet fails to penetrate it. Under normal conditions, however, the fabric would be as flexible as normal clothing.

So if you ever find yourself in quicksand, remember that gentle, sedate movements (like a slow breaststroke), not kicking legs and arms, provide the route to safety.

13 APRIL

How do flies crash into windows uninjured?

WHILE THAT HOUSEFLY on a collision course may appear to be zipping along, this is an illusion caused by its small size: its top speed is a modest 2 metres per second. For the blowfly it is 2.5 metres per second. Weighing 12 and 50 milligrams, respectively, the force with which the two hit the pane will be at least six orders of magnitude less than that of a 70-kilogram human walking into patio doors at 5 kilometres per hour (or 1.4 metres per second).

Kinetic energy imparted on impact is proportional to mass and velocity squared, so we find that the human headbutts the glass with an energy between 400,000 and 3,000,000 times as much as the flies' bodies. To impart the same energy as a walking human, let alone a runner at full pelt, the insects would have to be doing between roughly 6000 and 12,000 kilometres per hour – or Mach 5 and 10, respectively.

Flies also have a stiff, tough outer skeleton of chitin in a protein-rich matrix. The chitin between the body segments and joints of the appendages allows the shock of impact to translate to movement of appendages about the joints. The fly is its own shock absorber.

Chitin is akin in its chemical and mechanical properties to the keratin of human fingernails, and the lack of injury to a fly upon hitting the pane might be likened to our ability to flick a paper pellet without harm to our fingernail or the quick beneath it.

Hitting a moving sheet of glass is a different matter. At modest highway speeds, a car windscreen travels more than 30 times faster than a dawdling fly. The resulting squared velocity inserted into the equation above means that the imparted energy is perhaps 1000 times as great. The fly is of negligible mass relative to the car, so essentially it absorbs all that energy and the windscreen bears a red streak of the fly's optic pigments.

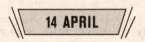

14 APRIL

Can iron spontaneously explode?

When the *Titanic* was being salvaged from four kilometres down on the seabed, great care had to be taken when bringing cast-iron objects to the surface, due to the risk that they might explode when they emerged from the water.

There are several phenomena involved. One is that cast iron

invariably contains small gas cavities or blowholes that are formed well beneath its surface. Another is that it has quite low ductility, and will fracture rather than deform. Thirdly, it is a very heterogeneous material, containing about 4.5 per cent carbon and significant amounts of silicon and manganese, together with phosphorus and sulphur. The principal phases that are present are graphite, argentite and ferrite.

When immersed in an electrolyte such as seawater, electrolytic corrosion starts up at the surface of the casting. One of the products of this corrosion is hydrogen in an ionic or atomic state. In this state it can diffuse through the ferrite lattice and find its way to the gas cavities. There it re-forms as molecular hydrogen, increasing the pressure in the cavities.

Because this electrolytic process takes place at great depth and pressure, the pressure build-up in the gas cavities reaches equilibrium with the external water pressure. Raising the cast-iron object from the deep seabed removes the external pressure on the iron, so the gas in the cavities creates very high stresses. At best, the iron will develop cracks. At worst, the casting will shatter.

15 APRIL

Why do birds sing at dawn?

AH, BIRDSONG. MUSIC to the ears of morning larks or, perhaps, an annoyance to those who prefer a quiet lie in.

The main function of most birdsong is long-distance communication, either to mark territory or to be sociable. As such it is largely intraspecific; blackbirds sing to impress blackbirds, not buntings. In contrast, social vocalisation, such as coordinating group activity, largely occurs at short range during active flight or foraging, or when settling down for the night or preparing to take flight as a flock.

Like any form of communication, birdsong bears an energy cost and requires channel capacity, which is limited largely by background noise and the quality of the medium, in this case air. In the mornings and evenings the air tends to be still, which reduces competing noise. It is also cooler at low altitudes, which favours transmission of clear sounds. Also, few birds forage at dusk, so in terms of energy use that is an economical time to perform.

Because much birdsong is territorial, it is practical for each species to sing at fixed times to avoid wasting energy on talking when no one is listening or when other species are competing for air time. Ideally, that male blackbird would be saying: 'If you are a male, keep off, or else! But if you are a female, let's get together.' Later, when other species are singing, he can go off to catch the early, deaf, worm.

16 APRIL

Why do thoughts feel as if they are in your head?

YOUR HEAD IS the focus for receiving information from your environment, mainly through sight and sound. Thought is secondary to sensory perception, but both have to combine quickly when you are threatened, and evolution has sited the organs for sensing information and processing it close together to ensure speedy response to dangers such as sabre-toothed tigers and muggers.

Researchers at the Karolinska Institute in Stockholm, Sweden, investigated out-of-body illusions using a brain scanner. They found that the brain's posterior cingulate cortex combines the feelings of where the self is located with that of body ownership. This is why you feel as if your thoughts are inside your head. Sensory deprivation can cause such feelings to be lost. When people lose sensory awareness while they are still conscious they become disorientated and may experience out-of-body sensations, feeling as though their thoughts are no longer inside their head. This can occur during certain types of torture or can be self-induced through meditation or drug use.

17 APRIL

Do all points on Earth get the same amount of sunlight?

THE SHORT ANSWER is that all points on the globe do receive the same amount of sunlight, because any portion of Earth that points towards the sun in summer will point away in winter, so the extra daylight in summer precisely cancels the lack of daylight in winter. However, this is true only for a simple spherical body in a circular orbit about the sun.

First, the orbit of Earth is not circular. According to Kepler's third law, Earth travels faster when it is closer to the sun than when it is further away. Because Earth is closest to the sun in early January, it moves fastest during the northern hemisphere's winter months. This can be seen by the fact that the time from the autumn to the spring equinox is about three days shorter than the time from the spring to the autumn equinox.

Because Earth has more daylight during the northern hemisphere's summer and spends more time on the 'summer side' of the equinoxes, the northern hemisphere receives slightly more daylight than the southern. This amounts to about 6 hours of additional daylight per year at 50° north, with higher latitudes receiving even more daylight.

The second major effect is the refraction of sunlight by the atmosphere. Because of this, we can still see the sun after it has sunk below the horizon. Typically, sunset appears to happen when the sun is actually about half a degree below the horizon. Along the equator, the difference amounts to about 4 minutes per day when daylight is more than 50 per cent of the day.

While this gives some extra daylight to everyone wherever they live, it gives more daylight in areas where the path of the sun makes a shallow angle with the horizon. The shallow angle means that it takes the sun longer to reach half a degree below the horizon. So higher latitudes (both north and south) get the most additional daylight. At 50° latitude, this is up to 8 minutes more per day when daylight is more than 50 per cent of the day – or about 36 hours per year on average.

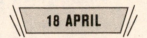

18 APRIL

How do plants know when to bloom?

Everyone knows that plants need light to grow and that most plants grow in different ways at different times of the year, particularly in temperate latitudes. This effect, called photoperiodism, enables temperate plants to control when they grow new leaves and when to drop them in the autumn.

However, it isn't the length of the day which is important, but the uninterrupted dark period – the length of night. During the period of darkness the plant produces a photosensitive protein called phytochrome, the concentration of which controls the onset of dormancy in autumn and bud-burst in spring. Even a small amount of electric light during the hours of darkness can upset this, and it is noticeable when a tree encroaches on a street light. In autumn, when the tree is starting to become dormant, the leaves closest to the light are often the last to fall.

Back in 1920, botanists Wightman Wells Garner and Harry Ardell Allard thought the length of daylight was the key to photoperiodism. So by the time it was discovered that the hours of unbroken darkness were responsible, plants had already been labelled as 'long-day' 'short-day' or 'day-neutral'. Chrysanthemums, for example, are short-day plants, and will only flower if there are enough hours of darkness. Growers of these plants used to keep their greenhouse lights on all night to delay flowering until it was discovered that just a brief burst of light would suffice.

Even seeds are armed with phytochrome, which allows them to use the direction of sunlight to sense which way is up (this ensures shoots and roots head off in the right direction). They can 'calculate' how deeply they are buried from the light intensity, and can even detect the presence of overhanging foliage and postpone germination.

19 APRIL

What is the ideal altitude for running?

ALTITUDE AFFECTS RUNNING in two opposing ways. The power needed to overcome air resistance varies approximately with the velocity cubed and in direct proportion to the density of the air. The important consequences of this are that air resistance is far more significant at high speeds, and that it can be reduced by going to altitude, where the air is less dense.

The other effect of thin air is that the runner receives less oxygen. In a sprint lasting less than 20 seconds, most of the energy comes from oxygen-independent glycolysis, in which the muscles break down carbohydrates without requiring large amounts of oxygen. This, combined with the runner's high speed, means that sprint times will be quicker at high altitude. Run for more than about a minute and the benefits of altitude are lost, because the runner depends more on aerobic respiration.

The 400 metres athletics event sits somewhere between a flat-out sprint and an aerobic distance race. A 1991 study in the *Journal of Applied Physiology* suggests that the ideal altitude for this event would be between 2,400 and 2,500 metres. Indeed, one of the most enduring records was set at the Mexico City 1968 Olympics by Lee Evans in the 400 metres. Mexico City is situated at about 2,250 metres above sea level. The air is less dense because of the reduced atmospheric pressure – 580 millimetres of mercury (mmHg) compared with 760 mmHg at sea level – so you can't make a decent cup of tea because the water boils at 92 °C as opposed to 100 °C, but you can run faster.

20 APRIL

Why does chopping onions make you cry?

ONIONS AND GARLIC both contain derivatives of sulphur-containing amino acids. When an onion is sliced, one of these compounds, S-1-propenylcysteine-sulphoxide, is decomposed by an enzyme to form the volatile propanethial S-oxide, which is the irritant or lacrimator.

Upon contact with water – in this case your eyes – the irritant hydrolyses to propanol, sulphuric acid and hydrogen sulphide. Tearfully, the eyes try to dilute the acid. However, it is these same sulphur compounds that form the nice aroma when onions are being cooked.

Tired of weeping over the chopping board? The most obvious way of avoiding the onion's irritant is to stand as far away from the onion as your arms will allow. Another way of reducing the amount of irritant reaching your eyes is to breathe through your mouth. This means that instead of creating a current of air which flows up to your nose and onwards to the eyes, carrying the irritant with it, the air is either directed into the lungs when breathing in or forced away from the face when breathing out.

In order to ensure that you breathe through your mouth, hold a metal spoon lightly between your teeth. There will be space for the air to enter and escape, and while our mouths are open we breathe preferentially through them, rather than our noses. You could also try simple goggles, though you might look a little silly.

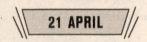

21 APRIL

Why do objects change colour as they get further away?

THIS EFFECT IS called aerial perspective and it is an important technique in a landscape painter's toolbox. Diluting colour intensity by blending it with white mimics the effect of the atmosphere on distant objects. It is the reason far-away hills have a bluish or purple tinge. In a landscape painting, bright colours such as reds and yellows are best used in the foreground, whereas pale blues and other diluted colours will give the illusion of depth. Of course, the Fauvists broke all these rules, favouring strong colours over realism, with wonderful results.

Another reason why the colour might change is that in the centre of your visual field there is a small area that is effectively blue-blind. This 'foveal tritanopia' means that objects with colours that differ only in how much blue they contain become indistinguishable when they are very small – about the size of a tennis ball viewed from the other end of the court. So white and yellow will look identical, as will red and magenta, or blue and black.

This phenomenon has been known empirically for a long time.

Naval signalling flags are designed so they cannot be confused even when viewed at a distance where this effect could manifest. Similarly, the rule of tincture in heraldry forbids a yellow emblem on a white background or vice versa, or blue on black, and so forth.

22 APRIL

How big is the biggest possible raindrop?

Molecules inside a water drop are pulled equally in all directions by neighbouring molecules. The attraction between molecules means the liquid squeezes itself until it has the lowest surface area possible: a sphere. The outer molecules are held by attractions within the drop and from the molecules beside them, providing the surface tension that allows insects to walk on water.

As a drop grows, this weak surface tension becomes increasingly unable to hold the spherical shape and the drop becomes distorted. This is why raindrops smaller than about 2 millimetres across remain spherical but, as they get bigger, they start to resemble a hamburger bun, with a flat bottom and a rounded top. Air resistance buffets raindrops as they fall, and water does not have sufficient viscosity to damp these disturbances so falling drops usually break up at a diameter of about 5 millimetres.

That said, raindrops between 8.8 and 10 millimetres were recorded twice a few years ago. The first time was by a research plane that flew over Brazil through cumulus congestus clouds that form over atmospheric regions undergoing convection – they are often created by strong updraughts. The drops are believed to have been formed by water condensing on smoke particles from burning forests below. Large drops were also spotted falling through clean, marine air over the Marshall Islands in Micronesia. They had formed within narrow regions of cloud where raindrops were able to collide and accumulate. It is unlikely that any of these large raindrops would have reached the ground intact because they readily break up as they fall, thanks to air resistance.

23 APRIL

Why do airplanes have round windows?

THE WINDOWS ON aircraft are small and round to make them safe. The first major jet airliner, the de Havilland Comet, had large, rectangular picture windows through which the passengers had a great all-round view. But after a few years in service, the aircraft started to break up during flight.

To find out why, de Havilland put a new Comet into a tank of water and then pressurised and depressurised it repeatedly to simulate the conditions of flight. After the equivalent of two years' worth of pressurisation cycles (which actually only took a few weeks in the water tank), the airframe was found to fail in the top corner of one of the large windows, which caused a catastrophic break-up in flight. The windows had to be redesigned and small, round windows set low in the fuselage were created. This solved the problem and the position of the windows remains the same today.

The problem was caused by stress concentration. You can see this for yourself with a bit of tape and some scissors. If you try to tear a piece of sticky tape across its width, it can be resistant. But if you nick it with a sharp object first it tears easily. This is because when a material has a crack in it, any stress concentrates around the crack tip. The sharper the crack the more stress becomes concentrated at the tip. Hence, even under small loads a crack can propagate through the material. If you punch a hole in the middle of the tape, however, you'll notice that it is harder to tear – the round hole distributes the force evenly.

24 APRIL

Can you catch up on missed sleep?

THE NEED TO sleep comes from a two-tier system. On one side there is sleep drive, also known as sleep pressure: the longer we are awake, the more this builds up, prompting our desire for sleep. This is the increasingly desperate need to sleep you might feel if you try to pull an all-nighter. As soon as you drop off, this sleep drive begins to dissipate.

Yet we don't feel more and more tired from the moment we wake up. For instance, people often feel more awake at eight in the evening than at four in the afternoon. That's because our sleep/wake patterns are also coordinated by the circadian clock in the brain. This produces what are known as alerting signals, which grow stronger for much of the day, to counter sleep drive. As bedtime approaches, these alerting signals drop off, and sleep drive overcomes us.

While we slumber, sleep drive diminishes until we can wake up feeling refreshed. Ideally, you shouldn't need an alarm clock to wake up in the morning – if you do, it's a sign you are accumulating a sleep debt. This will leave you feeling more and more tired as you fall into arrears. And the negative effects of getting too little sleep have been well documented – from memory loss to heart disease, weight gain and stroke.

So the simple solution is to repay this debt as soon as you can. In one study, researchers followed students who slept just four hours a night for six nights in a row. They developed insulin resistance (a precursor to type 2 diabetes), higher blood pressure and a rise in the stress hormone cortisol, as well as producing half the normal number of antibodies to this hormone. But all the effects were reversed when the students then caught up on the hours of sleep they had lost. Even so, not everyone is convinced that this kind of short-term recovery will alleviate the long-term health problems that result from regularly skipping sleep.

And there's another catch that might keep you awake at night. We know that shift work and jet lag, which play havoc with your body clock, also impair your health. Regularly sleeping at the wrong time can lead to diabetes, obesity and cancer, among other problems. And it seems that catching up on missed sleep at the weekend, a phenomenon known as social jet lag, can cause the same kinds of health problems as shift work. The healthiest solution is an early night.

25 APRIL

Are potatoes poisonous?

You may have been told as a child to never eat the green areas of skin on an old or damaged potato – but why is that? The green colouring is, in fact, harmless chlorophyll, but it acts as a warning that the potato has an elevated level of solanine. The entire potato should be discarded. The same applies to potatoes that have begun to sprout and to potatoes that show black streaks from late blight.

Potatoes are a member of the Solanaceae family of plants that include tomatoes, peppers, aubergines, tobacco and deadly nightshade. They are characterised by their ability to produce toxic alkaloids such as solanine in their leaves, roots and fruit. Unfortunately, solanine does not dissolve so it cannot be removed by soaking the potato and it is not destroyed by cooking.

When a potato is exposed to light, its solanine content escalates as a natural protection against being eaten by foraging animals. It is, after all, meant to propagate a new plant rather than be consumed. Solanine gives potatoes a bitter taste and checks the action of the neurotransmitter acetylcholine. This causes dry mouth, thirst and palpitations. At higher doses it can cause delirium, hallucinations and paralysis.

Even in edible varieties, a high concentration of glycoalkaloids is present in the leaves, shoots and fruit and these should never be eaten. Potato tubers should always be stored where it is cool, dry and dark, because those exposed to light may develop unacceptable concentrations of glycoalkaloids, indicated by the green colour you find in ageing potatoes.

It is sometimes said that if the potato were introduced to Europe today, instead of in the 16th century, the European Union would have banned it under the Novel Foods Regulation (EC) 258197. This requires all foods that do not have a history of consumption in the EU before May 1997 to undergo a pre-market safety check. The importance of checking was dramatically demonstrated in the US where a man almost died from eating the Lenape variety, introduced in 1964 without screening for glycoalkaloids.

26 APRIL

Why does skin wrinkle in water?

IF YOU'VE GONE swimming, or had a long bath, it's not long before your fingers start doing prune impressions. So what's going on?

The tips of fingers and toes are covered by a tough, thick layer of skin which, when soaked for a prolonged period, absorbs water and expands. However, there is no room for this expansion on fingers and toes, so the skin buckles.

Your whole body does not become crinkled as the skin has a layer of waterproof keratin on the surface, preventing both water loss and uptake. On the hands and feet, especially at the toes and fingers, this layer of keratin is continually worn away by friction. Water can then penetrate these cells by osmosis and cause them to become turgid.

A study in 2011 showed that wrinkles form a pattern of channels that divert water away from the fingertip – akin to rain treads on tyres. Scientists think that it could be an adaptation that gives us better grip underwater.

Could we record the past on a camera placed a light year away?

IF A CAMERA was placed 1 light year away from Earth with a high enough definition, could it be used to spy on events that took place on Earth one year ago? And, if so, could this technique be used to record our past by sending an array of such cameras to the appropriate distance in order to capture momentous events in Earth's history?

It's a cool idea, and in theory, yes, a camera placed 1 light year away could indeed record events on Earth a year earlier. Of course, because it would need to transmit its information back to Earth via radio waves, which also travel at light speed, it would take another year for the images to be returned.

The real problem would be positioning the camera at that distance. The fastest you could move the camera to that position would be at some fraction of the speed of light. Even if you could send your camera out at light speed, the camera will never 'catch up' to light that left Earth before the camera did, so the camera can only capture images of events which happen after it is launched. In other words, it amounts to an elaborate and expensive mechanism for recording the present, but in a way that means you can't view the results until a year later. You could record the present much more easily with an earthbound camera, and wait a year to view it.

Although we can't outpace and photograph the light that left Earth at any time in the past, if we located a black hole, say 50 light years away, and if its surroundings were perfectly free of obscuring gas and dust, then in principle we could see light from Earth bent around the hole. This is because black holes have the gravitational strength to bend light through 180 degrees – and direct it back towards us, giving us a view of events 100 years in the past. Achieving this might be a bit challenging, though.

Another challenge would be that building a camera with the necessary resolution would be no mean feat. Creating a sensor with enough megapixels, or a roll of film with fine enough grain would be hard, but the trickiest part would be making the lens. The maximum resolution a lens can attain is limited by its size. To recognise objects a centimetre across, from a light year away, you would need a lens several hundred times the size of the solar system.

28 APRIL

Why should you always cut spinach with a stainless steel knife?

THE REASON YOU should you always use a stainless steel knife to cut your spinach is intriguing and is a major stumbling block to the fortification of food with iron. Remember that a lack of iron is the world's most prevalent nutritional deficiency.

If you cut spinach with a normal iron kitchen knife, both iron blade and spinach will become discoloured because of the reaction between the polyphenols in the spinach, and the iron blade. If you want to see a dramatic illustration of this, make yourself a cup of tea and add a few crystals of a soluble iron salt such a ferrous sulphate (don't drink it).

The black discoloration you see is caused by the reaction between polyphenols, called tannins, in the tea, and the iron. The resulting black compound is highly insoluble. The implications for iron absorption in the body are huge because iron in this form is virtually unavailable for absorption. So whatever the source of strength in Popeye's spinach, it is not the iron.

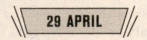

29 APRIL

Why isn't Pluto a planet any more?

To UNDERSTAND WHY Pluto got kicked out of the planet club, we need to roll back to the very beginnings of astronomy.

The word planet comes from the ancient Greek term *asteres planetai*, or 'wandering stars'. Even before the invention of the telescope, early civilisations noted there were seven astronomical objects that didn't remain fixed in the cosmos: the sun, the moon, Mercury, Venus, Mars, Jupiter and Saturn. With the Copernican revolution, astronomers realised we revolved around the sun, not vice versa, and swapped out the sun and the moon for Earth, giving us six wanderers.

Uranus was the first new planet discovered with the aid of the telescope in 1781, followed by Ceres, Pallas, Juno and Vesta. If you're not familiar with these, that's because we now call them asteroids – they are all small, mostly lumpy space rocks that orbit between

Mars and Jupiter – but for the first half of the nineteenth century, they were ranked alongside the larger planets, for a total of eleven. The discovery of Neptune in 1846, along with dozens more small bodies, forced a rethink, as it became clear these asteroids were altogether different.

Enter Pluto. As soon as it was discovered in 1930, astronomers could see it was an oddball world, orbiting at an angle to the other planets and occasionally crossing in front of Neptune's orbit. But Pluto was welcomed to planethood with open arms. The solar system was complete.

Or so it was thought. In the 1990s Pluto's status wobbled as astronomers began discovering trans-Neptunian objects (TNOs), a group of smaller bodies in a similarly distant locale, known as the Kuiper belt. Then in 2005 came the crushing blow: the discovery of Eris, a TNO that seemed to be a hair larger than Pluto.

Astronomers faced a crisis. Should Eris be anointed as the solar system's tenth planet? What if other would-be planets were lurking out there, ready to tear up textbooks at a moment's notice? The IAU was forced to react, and at a meeting in August 2006 it demoted Pluto to dwarf planet status.

Planets and dwarf planets are both round objects that orbit the sun, but dwarfs lose out on full rank because they have not hoovered up or displaced other smaller bodies from the neighbourhood around their orbit. Pluto falls foul of this clause due to other nearby TNOs, but Ceres gets an upgrade from mere asteroid to dwarf planet, since it is the only round object in the asteroid belt. The dwarf planet definition also explicitly excludes moons, otherwise our own moon, along with many others in the solar system, would qualify.

Does the moon change your weight?

TIDES ARE AFFECTED by the moon's gravity – and so are we. The weight of any object anywhere in the universe is the sum of the attraction between it and every other body in the universe, according to their various masses and positions.

This greatly complicates all sorts of measurements. For instance, in the eighteenth century the French astronomer Nicolas Louis de Lacaille went to South Africa and made sophisticated measurements of the shape of Earth's southern hemisphere, concluding it was flattened. He did not realise that he should have made these measurements on a wide open plain. His various readings were badly distorted by the gravitational masses of Table Mountain and the Piketberg range, and it was some time before anyone corrected the shape he calculated.

As the moon passes above, to the west, below and to the east of us, we get lighter, lean to one side, get heavier, then lean to the other side. However, its mass is so far away that if you were to swing a pin on a silk thread, this would affect your apparent weight more strongly than the moon does.

MAY

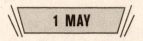

1 MAY

Can you measure the speed of sound at home?

You CAN INDEED, although the experiment is a bit low-tech and you'll need access to a very large space, such as a beach or a park, and another person's help . . .

Ask your helper to begin striking a hard (hammer-resistant) object once every second with a hammer – they'll need to use a stopwatch and maintain a regular beat. Begin walking away from your helper, looking back from time to time. When you are a few hundred metres apart, use a pair of binoculars if necessary.

As the distance between you and your helper increases, the delay between them striking the hard object and the sound reaching your ears will become greater. Eventually, the delay will match the time between each beat and the sound will once more appear to coincide with the action of the hammer. Sound has a relatively low speed in air, travelling at 344 m per second in dry air at 20 °C. This may vary at different temperatures, but not by much, so this experiment works pretty well just about anywhere.

When the sound of the hammer once again coincides with the 1 second beat, you are ready to measure the speed of sound. Stop walking away at this point and measure the distance between you and the helper. You should be 344 m away, or very close to that.

If outside noise is interfering with you detecting the beat of the hammer, ask your helper to increase the frequency of the beats to once every half second (in which case you should find yourself about

172 m from your helper when the beats again coincide with the sound), or once every quarter second if they can (when you'll be about 86 m away).

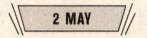

Why do we use toilet paper?

WHEN YOU STOP and think about it, it seems odd that humans are the only animals who see toilet paper as a necessity for dealing with our waste.

Although we share most of our DNA with great apes, there are some striking anatomical differences between ourselves and our nearest relatives, most notably our vertical posture. This enables us to walk tall with our hands free, but it also comes at a price: we experience problems with our back and joints, and the whole business of evacuating our waste is more difficult. The fundamental problem is that the area used for releasing urine and faeces is compressed between thighs and buttocks, so we are more likely than other animals to foul ourselves.

We also differ from other animals in our response to our waste, which we tend to regard with disgust. This seems to have developed as a result of living together in settlements rather than roaming through the forest, where we could leave our mess behind us. Unlike other primates we can learn when and where it is acceptable to excrete. Disgust is a sensible response to the threat of pathogens in human waste, and civilisation itself would be impossible without some system of sanitation. We have house-trained ourselves to the point where we are repelled by the very smell of our waste, and thorough cleansing has become a necessity for social reasons as much as for hygiene.

Wild animals, especially carnivores whose faecal matter contains material attractive to pathogens, have evolved to be able to clean themselves. You only have to watch cats 'playing the cello', as it is colloquially called, to see how proficient they are at grooming their rear. Parents will clean their young until they are supple enough to do it themselves. Adult animals will also groom each other, forming social bonds at the same time.

Human ingenuity in this respect has now gone far beyond toilet paper and wet wipes. In Japan there are toilets which will wash and blow-dry your most delicate areas without any effort on your part. On the other hand, the 16th-century French writer François Rabelais recommended using the softly feathered neck of a live goose for the ultimate in cleanliness and comfort.

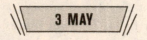

3 MAY

Why are pears pear-shaped?

APPLES, PEARS, MEDLARS, quinces and the fruits of related plants such as Pyracantha are known as pomes. The fleshy part of the pome grows from tissue between the stem and the carpels, the female reproductive parts of the flower.

Plant hormones called auxins, whose distribution is genetically controlled, guide all plant tissue growth and form, including fruit shape. As long as the final product does not put the plant at a disadvantage, selection has no influence on the fruit's appearance.

So what purpose or function does the shape of a pear serve? The answer is probably none. Indeed, in the wild the shapes of some varieties of apples and pears are almost interchangeable. As for the ones that are not, it is possible to speculate as to why. Some primeval pears might have benefited from having their fruit hanging out from between the leaves, or perhaps developing a longer neck made pears – which are softer and larger than, say, crab apples – less likely to drop prematurely. Or possibly some ancient seed-dispersing birds or bats were able to carry the necked fruit over longer distances.

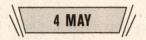

4 MAY

Why can bright light make you sneeze?

BECAUSE PHOTONS GET up your nose! Just kidding.

The answer may be fairly simple: when the sun hits a given area, particularly one shielded or enclosed in glass, there is a marked rise in local temperature. This results in warming of the air and a subsequent upward movement of the air and, with it, many millions of particles of dust and hair fibres. These particles quite literally get up one's nose within seconds of being elevated, hence the sneezing.

There has also been speculation that only some people sneeze in bright light, which suggests that the behaviour is genetic – termed a 'photic sneeze'. It seems to affect 18 to 35 per cent of the population. The sneeze is thought to occur because the protective reflexes of the eyes (in this case on encountering bright light) and nose are closely linked. Likewise, when we sneeze our eyes close and also water. The photic sneeze is well known as a hazard to pilots of combat planes, especially when they turn towards the sun or are exposed to flares from anti-aircraft fire at night.

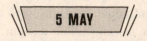

5 MAY

Why does athlete's foot often occur between the third and fourth toes?

THE ORGANISM RESPONSIBLE for tinea (better known as ring-worm), or athlete's foot, does not have an intelligent territorial instinct that leads it to a predestined home. Infections by the fungus responsible, *Trichophyton mentagrophytes*, begin in the space between the third and fourth toes because this location offers the ideal environment of a warm, dark and moist collection of dead skin cells.

The outer edges of the human foot are relatively flexible, having joints capable of motion in three planes. The spaces between all the other toes are therefore subject to a greater diversity of movement and forces, providing ventilation and the opportunity to slough off dead skin cells. Meanwhile, in the dark recesses between the less mobile third and fourth toes, a virtual agar plate awaits the arrival of *T. mentagrophytes*.

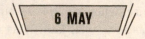

6 MAY

Is it possible to stop the Earth from spinning?

HOW MUCH FORCE would be required to stop the world spinning? If you used, for example, the engines of the space shuttle to do it, how long would it take? And what would be the effect on the planet, in particular the weather and the tides?

This is an excellent question for practising numeracy. All that is needed is some basic mechanics, made a tad more difficult for being rotational.

Rounding up some of the figures: the mass of the Earth (M) is 6×10^{24} kilograms and its radius (R) is 6.6×10^6 metres. Assuming it to be a solid homogeneous sphere, its moment of inertia (J) is given by $0.4 \times M \times R2$. It works out at 1×10^{38} kg m^2.

The planet spins once in 24 hours (86,400 seconds) so its angular velocity (ω) is 4.16×10^{-3} degrees per second, or, more properly, 7×10^{-5} radians per second.

Earth's angular momentum (h) is the product of the moment of inertia and angular velocity ($J \times \omega$), which gives 7×10^{33} newton metre (Nm) seconds. This is the momentum the shuttle engines will have to counter.

The thrust (F) of the shuttle engines on take-off is around 4×10^7 newtons and, if acting tangentially at the surface of the Earth, the torque (T) – or rotational force – about the Earth's centre is $F \times R$, which gives 3×10^{14} Nm. This torque acting over time (t) will change the Earth's angular momentum by an amount $T \times t$. The time needed to reduce it to zero is h/T or 3×10^{19} seconds, or 840 billion years. This is some 60 times the age of the universe, and by the time the shuttle had done its job there would be no weather or tides worth having. There is one other wrinkle: if the fuel needed comes from the Earth, the planet will get lighter and lighter. The whole of the Earth's mass will be expended as fuel long before the Earth stops spinning.

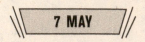
Does hay really spontaneously combust?

CUT HAY MAY look dry and lifeless, but it is considered inert only once its moisture content falls below about 15 per cent. When moisture levels are above 30 per cent, the tissues continue to respire, generating heat and water, which emerges as vapour through the leaf pores. Within the confines of the bale, the water condenses and spreads by capillary action, promoting bacterial and fungal growth, which adds respiring biomass.

In recently harvested hay, the result can be a single temperature peak of up to 60 °C between five and seven days from the day this process begins. This is self-limiting because the temperature kills most microorganisms and drives off the moisture. Sometimes a few weeks of warming cycles can follow as colonies of fungi wax and wane, but the successive temperature peaks likewise dwindle and the bale cools to match its surroundings.

However, if the hay has just been harvested, and the weather is humid, say, or wet with rain or dew, then a stack may sustain these warm conditions long enough to promote heat-loving microbes. These kick in at around 45 °C and die by 80 °C. They are not in themselves dangerous but as they raise the temperature, they can trigger exothermic chemical reactions that accelerate with rising temperature. As chemistry takes over from biology, temperatures can rise to a blistering 280 °C. Deep inside the bale this may stop at charring, but when the temperature rises above 231 °C the hay can auto-ignite on contact with air – it does not need a spark or a flame. The sequence varies with bale shape and size, porosity to air, and whether it is stacked or confined (in a barn, for example).

A dry outdoor bale of hay (which is grass and mingled herbs) or straw (which is the cereal stalks after the grain has been removed) could also conceivably catch fire from the sun's heat focusing on a discarded, half-empty glass bottle. Farmers know to monitor their stacks regularly.

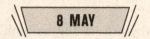

Why do loud noises hurt?

TWO FACTORS CONTRIBUTE to the unpleasantness of loud noises, like microphone feedback or a crash of glass bottles: the volume and the discordant frequencies. The human ear detects sound using tiny hair cells in the cochlea. A sound wave causes mechanical deformation of these cells sending an electrical signal to the brain via the auditory nerve. Different groups of hair cells react to different frequencies, with hairs detecting low notes at one end of the cochlear spiral, and high notes at the other. High volumes can impact the delicate hearing apparatus from chemical exhaustion of the cell-signalling process and mechanical damage.

We are adapted to find damaging stimuli unpleasant and painful so that we avoid them. Humans also find notes that are too close together very unpleasant – think of the discord between two adjacent notes on a piano compared with reasonably pleasant sounds obtained by playing notes spaced apart. This is probably because the closeness of the frequencies means there is an overlap in the groups of hairs activated by each note, overstimulating those that can sense both. Glass colliding as it is dumped into a truck creates many discords at once. For high notes like the crash of breaking glass the problem is worse because as the notes get higher the difference between adjacent notes gets smaller and more hairs will be stimulated by a given note. We are therefore more likely to experience discordance at higher frequencies – so don't feel guilty if the sound of a baby wailing makes you want to run for the hills.

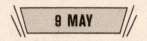

9 MAY

Why are flamingos pink?

THE EXPLANATION FOR why flamingos are pink can, surprisingly, be explained by answering another question first: 'why does a lobster turn red when cooked?'

Certain shellfish, such as lobsters, turn red when cooked because they are red to begin with – we just can't see it. In life, the red pigmentation in their shell, created by the presence of astaxanthin, is combined with a range of proteins and other pigments to produce the dull camouflage colours that enable the crustaceans to blend in with their environment.

When you boil a lobster, or a prawn or shrimp, proteins in the shell denature and unwind, releasing their attached pigments. But while the others break down at high temperatures, astaxanthin retains its stability and reflects light at the red end of the spectrum. This may seem ironic, given that astaxanthin belongs to the class of carotenoids known as the xanthophylls, which literally means 'yellow leaves'. Astaxanthin provides not only the colour for cooked crustaceans but also for red salmon.

Astaxanthin is used as a food additive for farmed salmon to make their flesh resemble the line-caught variety, and is also fed to intensively reared hens to give their egg yolks a richer orange colour. This practice has aroused controversy because it is felt that consumers might be misled into thinking they are eating an organic, hence arguably 'healthier', product.

Although there is no evidence that battery hens' feathers turn pink after eating astaxanthin, the pigment can make its way through the food chain into certain birds' plumage. The most famous example of this is the flamingo, which can obtain its rosy hue from astaxanthin in the shells of the shrimps that it sieves from brine pools.

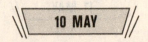

Why do boomerangs come back?

A BOOMERANG IS like two spinning aeroplane wings joined in the middle. It is held almost vertically before it is thrown end over end. Because it spins in this way, the top wing actually goes away from you faster than the bottom wing. This makes the sideways push on the top wing (similar to lift on an aeroplane wing) stronger than that on the bottom wing, so the boomerang gets tilted over, just as you would be if someone pushed on your shoulder, and its flight pattern begins to curve. Similarly, if you ride a bicycle and lean over, the bicycle will turn, eventually going in a circle. The boomerang does too.

Yet, most boomerangs don't come back and were never intended to do so. The Australian Aboriginal people made the boomerang for hunting and fighting rather than for sport or play, so they did not make the so-called returning boomerang throughout most of the Australian continent. For them, the real returns of boomerang throwing came in the form of fresh food or the beating of an enemy. The Warlpiri people can throw a karli boomerang and hit a target well over 100 metres away. Particularly skilled users of the karli throw this deadly weapon with surprising ease. The Warlpiri also manufacture the wirlki (also known as the 'hooked' or 'Number 7' boomerang), which is used for fighting. Across Australia, even in those areas where the boomerang is not made, there is near universal use of paired boomerangs as rhythm instruments in ceremonial contexts. Such boomerangs are still traded for ritual use across thousands of kilometres.

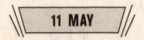

11 MAY

How do double-yolked eggs occur?

SOME SUPERMARKETS SELL eggs that are guaranteed to have double yolks – while this sounds as if it might need some extensive artificial engineering, the process is not as complex as it may sound.

The most common cause of double-yolked eggs is when two unfertilised egg cells are produced so close together that the hen's oviduct processes both yolks into the same shell. The tendency to produce more than one ovule at a time is influenced by genetics, so some breeds of chicken produce far more double yolks than others.

Certain cross-breeds ovulate rapidly, so that most eggs are double-yolked (and incidentally only rarely produce chicks). Some people prefer double-yolked eggs, regarding them as lucky, rather like four-leaf clovers. This is also reasonable from a nutritional point of view because most of the value of the egg is in the yolk, despite its high cholesterol level.

However, in some regions people prefer single yolks. This offers an incentive to market single or double-yolked eggs at a premium on the basis of regional tastes. This is aided by candling – you can shine a light through the eggshells to tell whether they are internally unusual, containing blood spots, double yolks or the like, and then sell them according to consumer preferences.

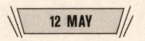

Why do some moles grow long hairs?

THERE IS MORE to hair growth than meets the eye. An eyelash, for example, is a marvel of programmed curvature, texture, colour, thickness, length and persistence. The hairs on a porcupine are a microscopic and macroscopic tour de force. Even the growth of simple human body hair requires elaborate control of tissues in the papilla at the base of the follicle from which any hair sprouts. Such control demands interaction with surrounding cells and the body's hormones.

If you have a long mole hair, it probably grows from papillae that originally produced fine, short vellus hair or humdrum body hair. The mole tissue grew from skin cells, mainly melanocytes, whose own growth controls had been disrupted by genetic or epigenetic changes to their DNA. They are now less inhibited and grow according to a programme differing from those of surrounding tissues, so their cells' textures and physiological products affect the growth of their neighbouring cells in turn. In particular, they can disrupt the activity of papillae, making them produce hairs unchecked. These hairs are often thicker and more simply structured than normal hair.

By contrast, a normal hair stops growing at shorter intervals, pauses for a suitable time, and is then shed after a few months or years of service.

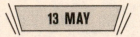
Which animals could beat an elite runner in a marathon?

SEVERAL LAND ANIMALS could beat a person over a marathon, including the husky, the camel, the pronghorn (a creature similar to the antelope) and the ostrich.

Accurate or not, the story of Pheidippides is the inspiration for the modern-day marathon. He is supposed to have covered 240 kilometres over two days in order to reach Sparta and summon aid when the Persians landed at Marathon in Greece. Later, he ran 40 kilometres from Marathon to Athens and used his final breath to announce Greek victory.

Some might argue that sending a messenger on horseback would have been more sensible. After all, a human has only ever beaten a horse twice in the 33 times that the Man versus Horse Marathon has been run in the Welsh town of Llanwrtyd Wells. However, this race is run over 22 miles (35 kilometres) and not the 26 miles and 385 yards (just over 42 kilometres) of the full Olympic marathon distance.

It is likely that long-distance running echoes a hunting strategy our ancestors developed more than 2 million years ago. Our prowess over distance relies on our ability to avoid overheating, accomplished mainly through sweating and being hairless. Cursorial hunters – those that are slower over short distances but have greater endurance – like humans simply have to run faster than the slowest gallop of a prey animal until it collapses with heatstroke.

Horses are significantly better than people over about half the marathon distance, which is why they were used for the Pony Express mail service in the US before the telegraph was introduced in 1862. Horses were ridden quickly for about an hour between stations, an average distance of 24 kilometres, and the riders would change horse at each station. Perhaps the Greeks used a runner and not a rider because a horse would not have coped with the heat of the late Greek summer and the mountainous terrain.

During the annual Iditarod trail dog sled race held in Alaska, the dogs pull the sled at around 24 kilometres per hour for up to 6 hours at a time. At these speeds, if they were running a marathon, Alaskan huskies would cross the finish line in less than an hour and a half – at least half an hour faster than the human world record.

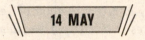

14 MAY

Why does garlic make your breath smell?

GARLIC PRODUCES A potent antifungal and antibacterial compound called allicin when the clove is cut or crushed. This is created by the enzyme alliinase acting on a compound called alliin. Allicin is responsible for the burning sensation you experience if you eat garlic raw.

However, allicin is not stable and generates numerous smelly sulphur-containing compounds, hence its pungent smell. After ingestion, allicin and its breakdown products enter the bloodstream through the digestive system and are free to leave again in exhaled air or through perspiration. This is the first effect of garlic.

In addition, the chemicals in garlic change the metabolism of the body and trigger degradation of fatty acids and cholesterol in the blood: this generates allyl methyl sulphide, dimethyl sulphide and acetone. These are all volatile and can be exhaled from the lungs, giving you garlic breath the morning after a meal. It is not necessary to eat garlic to have garlic breath because allicin can be absorbed through the skin. Just rubbing garlic on the surface of the body can be enough to generate smelly breath because it exits the body via the lungs.

The only real solution to smelly breath from garlic is for us all to eat it.

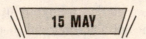

15 MAY

Why hasn't nature evolved wheels?

WHEELS ARE A pretty effective method of getting around, so why haven't more creatures evolved to have them? The fundamental obstacle is that evolution is not a process that 'thinks ahead'. It is merely the cumulative effect of natural selection acting on the results of chance mutations. Any new forms of life or locomotion only emerge if every single intermediate step conveys some sort of competitive advantage to the organism, or at the very least does not convey any disadvantage.

Therefore wings can evolve because a partial wing conveys some aerodynamic benefit to an animal leaping between tree branches. Similarly a hard shell can evolve because a partial shell conveys at least some protection. However, it is difficult to think of an intermediate stage to a wheel that would offer any competitive advantage instead of being an unwanted burden.

While we may not see any four-wheeled cats any time soon, there is one organism that has been using wheels to get around for millions of years: bacteria. It is the basis of the bacterial flagellum, which looks a bit like a corkscrew and which rotates continuously to drive the organism along. About half of all known bacteria have at least one flagellum.

Each is attached to a 'wheel' embedded in the cell membrane that rotates hundreds of times per second, driven by a tiny electric motor.

Electricity is generated by rapidly changing charges in a ring of proteins that is attached to the surrounding membrane. Positively charged hydrogen ions are pumped out from the cell surface using chemical energy. These then flow back in, completing the circuit and providing the power for the wheel to rotate.

It is a very sophisticated piece of nanotechnology and even has a reverse gear that helps the organism find food. So, far from nature not having invented the wheel, given the very large number of bacteria in existence, there are probably more wheels in the world than any other form of locomotion.

16 MAY

How do wind turbines work?

FOR GOOD REASONS, our electricity supply works not by providing a single, steady flow, but by flitting backwards and forwards. This is called alternating current, or AC for short. Older designs for wind turbines aim to keep the blades rotating at a constant speed to keep in time with this, because this enables optimal conversion of the movement into electrical energy for use in the national grid.

We convert the wind into electricity by allowing the turbine to spin freely and then 'braking' it to capture as much energy as possible. But instead of a mechanical brake such as those on a car or bicycle, which remove the energy from a spinning wheel, we use an electrical 'brake' that allows us to capture energy and convert it into electricity rather than waste it as heat. We don't want to stop the turbine completely, but allow it to turn at a constant speed regardless of how hard the wind is blowing.

If we want to freewheel a bicycle down a gentle hill at a steady speed, we apply the brakes gently and they warm up by the time we reach the bottom. If the hill becomes steeper, but we want to keep the same speed, we squeeze the brakes harder and they get even hotter because they convert more energy into heat (don't test this by touching the brakes).

Similarly, when we want our turbine to turn at 39 revolutions per minute in a steady breeze, we only apply a small amount of electrical braking, and generate a small amount of power as a result. When it's blowing a gale, we can really apply the 'brakes' hard to get a lot more power out, even though the blades are spinning at the same rate.

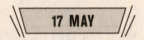

Why is beer packaged in brown bottles?

BEER IS USUALLY packaged in brown bottles to protect the drink's flavour being affected by sunlight. Nearly all beer contains hops. These provide the bitter taste and also act as a preservative (some natural deodorants even use hops for their antimicrobial action).

Hops contain isohumulones, which provide part of the bitter flavour – and this is where the problem with light arises. When UV light hits these compounds they decompose, leading to the creation of free radicals that react with sulphur-containing amino acids. The product is a thiol – the sulphur analogue of alcohols – and this leads to a 'skunky' flavour, so called for obvious reasons.

To protect against this, manufacturers use brown bottles that block out some of the UV rays. Brown bottles cut out most of the harmful wavelengths of light; clear and green bottles offer far less protection and so are more prone to develop a skunky flavour.

Try leaving a glass of beer out in the sun for even 10 minutes and compare it with one not in direct sunlight, and you will get an idea of what skunky beer tastes like.

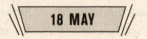

How do some animals change colour?

DID YOU KNOW that it isn't just the chameleon that can change its colour?

Many animals change the shade of their skin, and some can even change it to a different colour completely. Cuttlefish have pigment-filled sacs that they can control with their muscles, producing colour changes and patterns that are really amazing.

Many fish in the teleost group, such as the minnow, change colour in response to the overall reflectivity of their background. Light reaching their retina from above is compared in the brain to that reflected from the background below.

The interpretation is transmitted to the skin pigment cells via adrenergic nerves, which control pigment movement. Just like a printer has different ink cartridges, teleost skin contains pigment cells of different colours: melanophores (black), erythrophores (red), xanthophores (yellow) and iridiophores (iridescent).

Some flatfish, including flounder, go further than just reflecting what's around them. They develop skin patterns that make them even trickier to spot against the background. This seems to involve producing distinct areas of skin with predominantly, but not exclusively, one type of pigment cell. For example, black patches contain mainly melanophores and light patches mainly iridiophores, which can even produce a chequerboard appearance if the fish is lying over a chequered surface.

Would they still be able to change colour with a blindfold on? Since these responses are visual, blindfolding the fish would result in all the components of the chromatic system being stimulated equally. The fish would adopt an intermediate dark or grey skin tone similar to that on a dark night. And if the fish is receiving the same visual signals for a long time, the camouflaging effect will be longer lasting, hence the 'black' plaice sometimes sold in the UK, which have come from the sea around the dark volcanic seabed off Iceland.

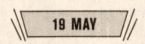

How does a yo-yo yo-yo?

INDEED, HOW DOES a yo-yo do lots of different things? There's no better way to find out than grabbing yourself a yo-yo and trying it out.

Play with a yo-yo in the normal way, making it travel up and down its string. You'll feel a slight jerk of tension when it reaches the bottom, before the yo-yo starts its journey back up in apparent defiance of gravity.

If you examine a yo-yo, you'll see it consists of two heavy discs connected by a smooth axle. The string is not tied to the axle but simply looped around it. You can visualise this string as a 'U' which starts at your finger, goes down and around the yo-yo's axle and back up again to your finger. The string is twisted, forming a loop at the bottom so that the yo-yo can't jump out of the 'U'.

As the yo-yo is released from your hand, its initial potential energy is converted into angular kinetic energy. It stores most of this energy, like a gyroscope, in its spin or angular momentum. The angular momentum of a rotating object is the measure of the extent to which the object will continue to rotate about its axis unless acted upon by an external force. A good yo-yo has more mass on the outside of the disc than on the inside in order to keep it light while at the same time maximising the angular momentum.

The end of the string loops loosely around the axle of the yo-yo, leaving the yo-yo free to spin inside the loop when it reaches the bottom of its descent. The friction between the string and the axle here is not great enough to allow the axle to grab the string and start winding itself up. However, if you give a small jerk to the string this momentarily reduces the string's tension. The friction of the string against the axle is then enough to allow the string to start wrapping itself around the axle.

Once the string has made one turn around the axle, the loop starts to act as if it is attached – the outer turns of string stop the inner loop from slipping any more and the yo-yo then begins to climb back up the string, converting its kinetic energy into potential energy.

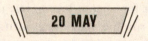

Is there a scientific way of scoring a goal?

Is THERE ANYTHING science can tell us about taking a penalty kick? Is there a well-founded, foolproof way of taking fear, emotion and human error out of the equation and guaranteeing a goal?

Unfortunately, there is no way to guarantee a goal. However, game theory suggests that if the behaviour of either the kicker or goalkeeper is predictable, then an advantage can be gained – not a 'guaranteed goal', but an increase in the percentage of successful shots (or, indeed, successful saves).

To prevent this advantage, the only solution is for both kicker and goalkeeper to choose a side at random – not easily done in the human brain. And goalkeepers may learn to read small movements made by some kickers.

In order to be truly random, perhaps coaches could flip a coin and then send an order to the kicker, and do the same for the goalkeeper. In this case, the kicker needn't try to kick to a spot that the goalkeeper can't reach, it being understood that 50 per cent of kicks will likely be blocked and 50 per cent will get through.

So much for the theory. Even if each game were totally random, there would still be a winner in a penalty shoot-out, who might well have many beliefs as to the 'reason' for success. You might as well ask a lottery winner the 'secret' of their winning.

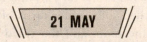

Can you stop cheese going mouldy?

MOULD IS A menace for cheese lovers, but there is a simple way you can stop it from ruining your lunch plans: put a lump of sugar under a cheese cover with the cheese and the cheese will remain mould-free.

So, how does it work? The sugar cube absorbs water, lowering the relative humidity, so that mould can no longer grow on the surface of the cheese. Salt would work just as well, as would saturated solutions of sugar or salt – saturated solutions are those that still contain some undissolved sugar or salt.

This forms the basis of humidity control in museum display cabinets. If the humidity is too high, undesirable moulds grow, but if it is too low, wood and leather might crack. Saturated solutions of different salts can peg the relative humidity anywhere from 10 per cent to 90 per cent. For example, a saturated solution of lithium chloride will maintain a relative humidity of 11 per cent, while a saturated solution of common salt keeps the relative humidity at around 70 per cent.

Is cracking your knuckles bad for you?

THERE ARE A couple of theories regarding what causes the 'pop' heard in knuckle cracking.

The common view is that it is caused by bubbles of gas within the synovial capsule of the metacarpophalangeal and phalangeal joints. The bubbles form when these joints are stressed – a process called cavitation – and then collapse as the pressure changes within the joint. It is the collapse that creates the noise. The energy released by this process has been estimated at just 0.07 millijoules per cubic millimetre. To cause damage to a joint, this figure would have to rise to about 1 millijoule per cubic millimetre. Cumulative damage from these pops cannot be ruled out, however.

The second theory is that the noise comes from the sudden deformation of the fibrous joint capsule itself and that the pop is its sudden slap onto the joint fluid within. This might cause microtrauma which could accumulate over years.

Regardless of the theories, there is little evidence that knuckle cracking causes arthritis: a survey of knuckle crackers showed no more incidence of arthritis than non-knuckle crackers. One American doctor went so far as to crack the knuckles on just one hand for 50 years to see if there was a difference between that hand and the other – there wasn't.

It is possible to cause acute trauma from the stress required to cause the joints to pop in the first place, of course, but one has to say that it's quite satisfying, isn't it?

Are bubbles always round?

BLOW A BUBBLE with some soapy liquid, and a neat sphere will lightly float into the air. But what do bubbles look like when other forces are at work, say, if they were blown underwater?

Bubbles can take on a number of different shapes depending on their size, velocity and the purity of the water. The boundary between air and water has surface tension, so a static bubble will take on a spherical shape to minimise its surface area. But as it moves through water, the influence of drag forces comes into play.

You might expect bubbles to adopt a streamlined aerofoil shape. However, the opposite is true. According to Bernoulli's principle, the surrounding fluid must accelerate as it passes the bubble, which results in a pressure drop around its equator. The bubble then expands so those larger than about 1 millimetre in size become noticeably disc-shaped as they rise.

As bubbles increase in size, turbulence causes them to oscillate, creating extra drag and slowing them down. Above about 20 milli-metres diameter the increased turbulence causes bubbles to enter what is officially called the spherical cap regime – a mushroom shape. The upper half of the bubble is a hemisphere, but the turbu-lent vortices at the rear cause the underside to become a chaotic mass of fine bubbles breaking and coalescing, and these get caught in the wake behind the rising spherical cap.

The pressure acting on a bubble decreases nearer the surface, hence bubbles expand as they rise, making them more likely to enter the spherical cap regime. Greater depth also provides a greater distance over which rising bubbles can collide and coalesce, as large faster-rising bubbles tend to catch up and engulf smaller ones above them. This process can continue until they reach the critical size for spherical cap behaviour.

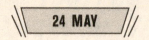

24 MAY

Could you survive on beer alone?

BEER HAS HAD a reputation since antiquity as being a staple in the diet, often called 'liquid bread'. In ancient Egypt, workers received beer as part of their salary, as did the ladies-in-waiting of Queen Elizabeth I of England. In 1492, one gallon of beer per day was the standard allocation for sailors in the navy of Henry VII.

This high reputation for beer came about because it was made from malted barley, which is rich in vitamins. This is still true today. A quick check using nutritional tables shows that a pint can provide more than 5 per cent of the daily recommended intake of several vitamins, such as B9, B6 and B2, although other vitamins such as A, C and D are lacking.

It is of course unethical to conduct an experiment to see whether one can live on beer alone. However, during the Seven Years War of 1756–63, John Clephane, physician to the English fleet, conducted a clinical trial. Three ships were sent from England to America. One – the *Grampus* – was supplied with plenty of beer, while the two control ships – the *Daedalus* and the *Tortoise* – had only the common allowance of spirits. After an unusually long voyage due to bad weather, Clephane reported that the *Daedalus* and *Tortoise* had 112 and 62 men respectively requiring hospitalisation. The *Grampus*, on the other hand, had only 13, arguably a clear-cut result.

Needless to say, the sailors' allowance of eight pints of beer per day is no longer within the accepted confines of current moderate alcohol consumption. One can only speculate on the state of their livers. Living on beer alone may be a fantasy for some, but it is not a good health strategy.

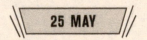

Where is water stored in the human body?

A 70-KILOGRAM MAN is 65 per cent water. That amounts to approximately 45 litres – so where does it all go?

Most of this is contained inside the trillions of body cells. A lesser amount surrounds these cells, while an even smaller amount circulates as plasma. Fluctuations in body weight over hours, days or a month are a feature of the body's ability to vary this volume. Variable tightness of finger rings, end-of-the-day aching legs and tight shoes are testimony to this movement of fluid.

Urine is the fluid kept in the bladder as a result of renal activity. The need to urinate is actually the result of only 200 to 300 millilitres of urine in the bladder. Faced with a bladder which is full and hasn't been emptied all day, the kidneys cut down on urine production. When the bladder is emptied at last, they work to restore the blood's correct electrolyte balance, and the bladder soon fills again.

Kidney activity is to some extent also controlled by the body's clock, and less urine is produced at night. Most people urinate several times during the day but will not have to get up during their sleep. If you reverse the cycle by flying halfway round the world you find yourself not needing to pee all day then getting up all night.

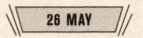
How can a plane fly upside down?

A PLANE FLYING upside down appears a baffling feat. The wings of a plane are designed to provide uplift when the plane is flying horizontally – so why doesn't the uplift, when a plane is flying on its back, work in reverse and force the plane back down towards the ground?

This is a common misconception. Although the aerofoil shape of an aircraft's wing produces some of the lift in normal flight, the more important factor is the angle of attack – the angle at which the air strikes the wing.

The wings of an aircraft are normally inclined to about 4° to the horizontal when compared to the main body of the aircraft. This is known as the chord angle of the wing.

So even when the fuselage is level, the angle of attack into the oncoming wind is 4°. This produces lift in the same way that your hand experiences an upward force when you hold it at about 45° to the horizontal in a fast-moving stream of air. Your hand does not have an aerofoil shape but the lift that you feel is caused by the angle of attack of your palm to the oncoming wind.

It is this principle that allows an aircraft to fly upside down. The nose is pointed further upwards than in standard flight because of the need to offset the chord angle of the wing. But if the angle of attack is positive compared to the relative airflow over the wing, then an upward force will still be produced. It is this lifting force which overcomes the force produced by the shape of the wing, and holds the aircraft in the air.

The bigger problem that pilots should be concerned about when flying their aircraft upside down is the risk of the engine stopping, because both the oil and fuel systems in most ordinary light aircraft are fed only by gravity. Flying your aircraft upside down can easily cut off the fuel supply because the valve that is feeding fuel to the engine suddenly finds itself at the top of the tank.

27 MAY

What would happen to a helium balloon released into space?

THE SIMPLE ANSWER is that the balloon would behave in exactly the same way as any other object released into space: it would continue to travel along the same trajectory until another force was applied to it. If it were released close to a planet, or similar object, then it would move into orbit around it.

This was what happened to one of the US's earliest experiments with communication satellites, Echo 1A and Echo 2 (Echo 1A is popularly known as Echo 1, but the original Echo 1 satellite was actually destroyed when its Delta launch vehicle failed on 13 May 1960).

The Echo satellites were both balloons (nicknamed 'satelloons' by NASA technicians), 30.5 metres in diameter when fully inflated, and constructed from a metallic-coated Mylar skin. The satellites were intended to reflect radio transmissions, particularly intercontinental telephone and TV signals. The Echos' mode of operation was entirely passive: radio waves simply bounced off their shiny surfaces, which, combined with their relatively low orbits (between 1,519 and 1,687 kilometres above the Earth), made them clearly visible across the entire globe.

Indeed, the satellite's remarkably high albedo, which made it seem brighter than a first-magnitude star (the brightest star in any given constellation), once caused considerable embarrassment to Arthur C. Clarke. At the time, he and Stanley Kubrick were discussing ideas for a movie script that would eventually become *2001: A Space Odyssey*. Clarke had just talked the director out of including UFOs in the plot synopsis when both men glanced at the sky over Kubrick's New York apartment and were stunned to spot a classic flying saucer passing serenely overhead. It was Echo 1A.

The Echo balloons were also involved in research into atmospheric density, solar pressures, the dynamics of large spacecraft and global geometric geodesy. Indeed, the Echo programme enabled the Pentagon to pinpoint the precise location of Moscow for the unfortunate purpose of accurately targeting the Soviet capital with its missiles. A third NASA satelloon, PAGEOS, intended purely for geodesic studies, was launched into a polar orbit in 1966.

Their success illustrates that a balloon can be safely inflated in space, although there are some distinctive aspects of operating balloons in space. If the envelope is exposed to direct sunlight, then the balloon will spin because the gas on the sunward side is expanding faster. It would also be subject to increased thrust from the gas escaping on that side: no material is entirely gas-proof when it comes to holding either hydrogen or helium, and the expanding gas would stretch the balloon's skin and make it more permeable as a result.

Given sufficient time, all the gas within the envelope would escape. The balloon would not, however, deflate in the accepted sense: instead it would retain its spherical form even when completely empty. This is because there is no external pressure to force it to collapse, as there is in the Earth's atmosphere.

Why are there only two sexes?

SOME SPECIES DO have more than two sexes. Single-celled ciliates have up to 100, and mushrooms have tens of thousands. But most organisms – even single-celled ones – come in two types.

So why are there two types in most species? In all species, no matter how many types, procreation requires just two cells and any can mate with any other sex cell that is different from it. But finding just two types in most species seems paradoxical, because having many types would maximise the chances of finding a mate.

One answer to this problem is that two types is best for coordinating the inheritance of cytoplasmic DNA – the part of the cell's genetic material that is not contained in the nucleus. However, there is a drawback to this solution. The species with two types fuse cells and potentially run the risk of scrambling this extra material.

The species with more than two mating types do it differently. With three types, the coordination is even more difficult to make error-proof, while those with many mating types don't fuse cells at all and so are not constrained to having just two types.

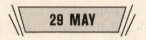

What is the highest-altitude parachute jump?

THE HIGHEST-ALTITUDE PARACHUTE jump was made by Joseph Kittinger of the US Air Force, who jumped from a balloon at 31,333 metres on 16 August 1960. He was in free fall for 4 minutes 36 seconds, reaching an estimated speed of 1,150 kilometres per hour. He opened his parachute at 5,500 metres.

Normally, jumps are made from below 4,200 metres, a limit set by the risk of anoxia (when your body or brain loses its oxygen supply). And in jumps from higher, the thickening air as the parachutist falls can create problems.

In the lower atmosphere a skydiver accelerates down for about 10 seconds until the increasing drag matches his or her weight, which happens at a terminal velocity of about 55 metres per second. As air thickens, terminal velocity decreases. For most free falls, skydivers are decelerating.

Falling through higher, thinner air, you would be travelling faster than the terminal velocity of the lower air when you meet it and the drag force peaks. Effectively, you would collide with the atmosphere. During his jump in 1960, Kittinger felt this force as a choking feeling, peaking at about 1.2 *g* at around 23,000 metres.

A fall from 75,000 metres would give a 3 *g* impact with the atmosphere at about 31,000 metres, which would wear out over 20 seconds or so, when the jump would become an uneventful skydive. A skydiver re-entering from low Earth orbit need not suffer much more than 3 *g* if they position their body across the airflow to lengthen the time of impact with the atmosphere, but it would get very hot.

Kittinger wore a full pressure suit to protect himself from the low pressure of the stratosphere. However, the main problem with this kind of jump is retaining stability during free fall. Kittinger's equipment included a small stabilising parachute, but this failed during his first attempt, putting him into a 120 revolutions-per-minute spin as he descended. He lost consciousness, only surviving because his automatic main parachute opened.

30 MAY

Can you tell how much water is in a cloud just by looking at it?

THE AMOUNT OF water inside a cloud is not different from the amount of water in the clear air around it. However, while the clear air contains water vapour, inside the cloud the air is saturated with water vapour and it has condensed out to produce the cloud. The difference between the two states is caused by temperature differences rather than the water content.

The colour of a cloud doesn't make much difference either. In the higher part of a cloud, the water is in the form of ice crystals. Lower down it is a mixture of ice and liquid water. The colour of the cloud depends mainly on this ice/water mixture and the size of the water droplets, and less on the total amount of water.

An estimate of a cloud's water content can be obtained from the amount of rain the cloud can produce. If the entire atmosphere were saturated with water and it all fell in a steady stream, this could produce some 35 millimetres of rain, while the thickest actual clouds produce about 20 millimetres. Cloudbursts can produce 50 millimetres or more, but this requires additional moisture from the surrounding atmosphere, which means such events are localised.

The heaviest cloudbursts roughly follow the following equation:

the rainfall in millimetres equals 6.5 multiplied by the square root of the time in minutes that it has been raining. A more typical shower produces a few millimetres of rain, at a rate of perhaps 0.1 millimetres per minute. Normally 1 millimetre of rain corresponds to 1000 cubic metres of water weighing a thousand tonnes per cubic kilometre of cloud, though the thickest clouds can contain up to 20 times as much.

You can also estimate the amount of water from the volume of the cloud. By volume, the fraction of the cloud filled with water is about 1 millionth, or 0.0001 per cent. The cross-section area of a cloud can be measured from its shadow. A small cloud 500 metres by 500 metres and 100 metres high has a volume of 25 million cubic metres, of which roughly 25 cubic metres will be water, weighing 25 tonnes. Even if you can't calculate the precise amount of water in a cloud, these numbers may still impress your friends.

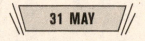

31 MAY

Why does a whip crack?

THE CRACK IS actually a sonic boom, caused by the end of the whip breaking the sound barrier. This is possible because a whip tapers from handle to tip. When the whip is used, the energy imparted to the handle sends a wave down the length of the whip. As this wave travels down the tapering whip it acts on a progressively smaller cross-section and a progressively smaller mass.

The energy of this wave is a function of mass and velocity and since the energy of the wave must be conserved, it follows that if mass is decreasing as the wave travels down the whip, then velocity must increase. Therefore, the wave travels faster and faster, until by the time it reaches the tip it has attained the speed of sound.

When the wave reaches the tip of the whip it must be dissipated. Some goes to the air and some into a reflected wave that travels back up the whip. At the point that the initial wave reaches the tip and is about to embark on its return it undergoes a brief but enormous acceleration. The result is that it moves supersonically.

JUNE

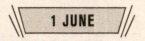

1 JUNE

How do ants know to gather in large numbers?

ANTS FORAGE RANDOMLY. On their way back to the nest with food they leave a pheromone trail that other ants will randomly encounter and identify as 'pointing' to a food source, according to the concentration of pheromones at each point.

The more food, the more pheromones, and the more ants will pick up the trail. It's a fascinating and simple system they have used for millions of years. Other ants follow the pheromone trails and also reinforce them, but once all the food is consumed, they stop leaving pheromone trails. Pheromones evaporate over time, preventing areas from becoming saturated and confusing. Many studies have modelled ant foraging behaviour; some are impenetrable, but others can make very interesting reading.

Ants are generally unwelcome in the house, and the best way to prevent them coming in is to follow a homeward-bound individual to its point of entry and dowse about a square metre around the area with boiling water, followed by a small amount of anti-ant powder.

Ants might not have the charming waggle dance of the bee, but that's not to say that they don't communicate with each other. In fact, they nod to each other when passing on a route, just like truck drivers do.

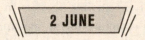

Why do bananas turn brown quicker in the fridge?

THIS IS COUNTERINTUITIVE to those of us brought up to believe that chilling foodstuffs slows decay, but a simple experiment will show us if it's true or not.

Place one banana in the fridge and leave the other at room temperature (approx. 20 °C). Over the course of a day, you'll see that the banana in the fridge will brown or blacken faster than the one at room temperature.

While many fruits are stabilised by refrigeration, most tropical and subtropical fruits, and bananas in particular, exhibit chill injury. Tests show that the ideal temperature for bananas is 13.3 °C. Below 10 °C spoilage is accelerated because their cells' internal membranes are damaged, releasing enzymes and other substances.

The membranes that keep separate the contents of the various compartments inside a cell are essentially two layers of slippery fat molecules or lipids. Chill these membranes and the molecules get more sticky, making the membranes less flexible. Bananas adjust the composition of their membranes to give the right degree of membrane fluidity for the temperature at which they normally grow. They do this by varying the amount of unsaturated fatty acids in the membrane lipids: the greater the level of unsaturated fatty acid, the more fluid the membrane at a given temperature. If you chill the fruit too much, areas of the membrane become too viscous and it loses its ability to keep the different cellular compartments separate. Enzymes and substrates that are normally kept apart therefore mix as the membranes collapse and hasten the softening of the fruit flesh.

Skin blackening involves the action of a different enzyme from those involved in flesh softening. In the skin, polyphenol oxidase breaks down naturally occurring phenols in the banana skin into substances similar in structure to the melanin found in suntanned human skin. So the browning starts sooner in refrigerated bananas because of chill-induced membrane damage that allows the normal process of decay – which would have occurred at room temperature – to begin earlier. The cold itself does not speed up the browning part of the reaction. Indeed, if chill damage in a fridge is induced first, removing the banana from the fridge then hastens the process as the reaction that causes the browning, once it is under way, is speeded up by heat.

This can be demonstrated by putting a banana skin in a freezer for a few hours. The inner surface will stay creamy white because, although the membranes are destroyed by the freezing process, the oxidases cannot work at such low temperatures. Then let it thaw overnight at room temperature. In the morning it will be pitch-black due to the damage the membranes suffered in the freezer. Yet had the cold itself caused the blackening, it would have turned dark while it was being frozen.

3 JUNE

How do our brains tune out background noise?

IF YOU TAPE a conversation in a restaurant and play it back, the background noise – like cutlery hitting plates, the chatter of other guests and doors closing – will be surprisingly prominent. Yet our brains can filter out these sounds as they happen. How?

Microphones are wonderfully objective devices. They detect variations in absolute pressure or the pressure gradient on a particular axis and faithfully transduce these into electrical signals. In contrast our ears have a brain attached, and between them they do a much more subjective job – interpreting our acoustic environment, not just recording it.

Our ears themselves are simple pressure transducers, but we also have ways of working out where sounds are coming from. For this we make use of the relative levels, phase and arrival times of sounds. In addition, the shape of our head distorts the local audio field in a way that we are personally familiar with, and this aids location of sound sources, particularly when we can move our heads.

We detect not only direct sounds but also reverberant ones. The space we are in significantly colours and adds to sounds, mostly in the form of delayed noise from random reflections. This would severely reduce the intelligibility of sound if the brain were not adept at adjusting to these conditions. It works out when the noises arrive, and where from, and can largely ignore them if it so chooses.

When the sound objectively recorded by a microphone is replayed through a single loudspeaker or even a stereophonic system, the random cutlery-clanging reverberant sound that should be all around us is now directly in front of us. Directional and timing cues that the brain would normally use for filtering are now inconsistent or just plain wrong.

4 JUNE

Are there any insects that live on or in the sea?

MARINE INSECTS CERTAINLY exist, although they constitute only a tiny percentage of the total number of insect species. The species range from those which live in brackish water, for example in saline lagoons and rock pools at the upper limit of the splash zone, to species which live on or beneath the sea surface.

A survey of brackish lagoons in Ireland recorded 77 species of insect, although most of these are not confined to the habitat. Flies (Diptera), beetles (Coleoptera) and true bugs (Heteroptera) were the commonest groups. The flies are usually present as larvae, the others in both the larval and adult stages.

True marine insects can be divided into species that live their entire lives on the sea surface and species that can tolerate immersion either as adults or as larvae. The five species of sea skaters (*Halobates*) are examples of the first group and are often cited as perhaps the most extreme example of marine insect. It is true they never voluntarily venture onto land, but they actually live only on the water surface without getting wet – just like pond skaters and water striders, their freshwater fellow members of the family Gerridae. *Halobates* lay their eggs on flotsam and jetsam. They can be found in all the world's oceans between the latitudes of around 40 degrees north and south.

Arguably, species such as the true bug *Aepophilus bonnairei*, an inhabitant of rocky Atlantic shores of western Europe and north Africa, are more truly marine because they can survive regular and prolonged immersion in seawater. At high tide, adult and larval *Aepophilus* shelter in rock crevices or in the holes drilled by rock-boring molluscs. They emerge at low tide to feed, searching for prey in seaweed. *Aepophilus* tends to be most common on the lowest part of the shore and therefore spends most of the tidal cycle under water. Consequently, it is more often seen by marine biologists than by entomologists.

How much of the Sahara is sand?

AT MORE THAN 9 million square kilometres, the Sahara is the world's largest desert, but it has no uniform topography. About 15 per cent of the area is covered by sand dunes, 70 per cent is made up of stone deserts of denuded rock and coarse gravel, while the rest is oases and mountain ranges.

Beneath the sand dunes lies rock of varying types. In Algeria and Libya, deposits of oil and gas have also been discovered, although inaccessibility has hampered their exploitation.

Not all the Sahara is sandy, but wherever wind-blown sand settles for any reason, more is likely to collect in the same place. In these places, sand accordingly forms wind-driven dunes, whose behaviour is astonishingly complex.

Surface sand seldom piles higher than a couple of hundred metres above underlying earth or bedrock, except by filling ancient valleys or lakes. Such deep sands and spongy sandstone form important groundwater reservoirs. And yes, sand does form and re-form constantly as water erosion, frost and wind-driven particles flake grains off rocks. Conversely, deep moist layers of sand become cemented into sandstone, which in turn may go through the same cycle after millions of years.

Far deeper sand occurs in submarine detritus fans, which build up at the mouths of rivers. Even more intriguingly, the Mediterranean has dried up repeatedly in the last tens of millions of years. Each time that happened, rivers flowing into the basin eroded their fans into massive canyons, which silted up again whenever the sea returned. The bed of the lower reaches of the Nile consists of silt, more or less compacted, kilometres deep. The river's erosive strength can turn this into an underworld canyon that has dwarfed the Grand Canyon in the past.

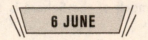

6 JUNE

Could we build technology to change the galaxy?

PREDICTING THE FAR future is a fool's game. So let's take the usual dodge and say: unless something is forbidden by known physics, it will be done. Eventually. Before we start, let's invent two things: self-repairing AI supervisors that can direct projects lasting many millennia; and vehicles that can reach close to the speed of light, maybe riding on laser beams or driven by miniature black holes.

Thus tooled up, we could hop from solar system to solar system, sweeping across the galaxy in 10 million years or so, and then out into our local supercluster of galaxies. The potential building site is quite spacious. Such a civilisation would consume lots of energy, and that is where our engineering may be most conspicuous. One option will be to plug in to local power sources, such as harnessing starlight with orbiting solar power stations. As demand for power grows, lots of these could be arranged to obscure a star completely, forming a closed 'Dyson sphere', named after physicist Freeman Dyson, who pointed out that technological civilisations tend to use ever-escalating quantities of energy.

If we built one we would darken the sun and leave a vast archaeological ruin in the event of our demise. Today, Earthly astronomers are looking for the darkness cast by alien engineering on this scale. With this level of technology we could even move stars around, albeit slowly. The simplest way would be to place a mirror on one side of the star to reflect some of its light into a beam, producing thrust in the opposite direction. Or the energy from a Dyson sphere could power ion engines to move the star a bit faster.

If wrangling mere stars seems humdrum, how about building the ultimate particle accelerator, capable of reaching the immense energy where all forces are unified and the fundamental nature of space-time is finally revealed. To be able to boost particles to the required energies this machine would have to stretch at least a hundred times the distance from the sun to Pluto.

If our hunger for power isn't satisfied by stars and supermassive black holes, then we might learn to create microscopic black holes and feed them with dust. This could unlock the mass-energy of inert matter, turning it into hot Hawking radiation that could be used to drive our interstellar industry. This could change the future of everything. With civilisation spreading through space at close to the speed of light, it would fill the cosmos with waste heat and so change its physical properties. The conversion of matter to radiation would even slow down the expansion of space a little, which puts our petty meddling with Earth into perspective.

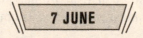

7 JUNE

What is the length of the longest straw you could still drink from?

IF YOU APPLIED an absolute vacuum above a non-volatile liquid, then the maximum height you could suck it up a vertical pipe would be reached when the hydrostatic head pressure of the column of liquid equals one atmosphere (101,325 pascals). This pressure is given by p × g × h, where p is the fluid density, g is the gravitational acceleration (9.81 metres per second per second) and h is height in metres.

For water, which has a density of 1000 kilograms per cubic metre, this gives a maximum height of about 10.3 metres.

However, because water has a vapour pressure of 3536 pascals (at 27 °C), it will begin to boil before you reach a perfect vacuum. So the maximum vacuum pressure that you could apply is 101,325 - 3536 = 97,789 pascals. This gives a maximum height of 9.97 metres.

In the case of a soft drink, things are more complicated, because the dissolved carbon dioxide will start to 'boil' out of solution under vacuum. If you sucked extremely slowly, first of all you would only get CO_2 and then, when you had removed the gas, you would get flat soft drink. If you sucked very quickly, then you might get the drink to rise up before the CO_2 nucleated and formed bubbles. More likely you would get a froth of liquid and CO_2 bubbles, and you might actually be able to suck this up to a much greater height because the effective density of the foamy mixture would be lower than pure liquid water. At intermediate suction rates, the foam bubbles would coalesce and you would be limited to a lower column height.

The exact answer depends on how much dissolved CO_2 you want left in your drink and the maximum rate at which you can suck. You would also need more than an ordinary drinking straw, because plastic ones collapse under moderate vacuum pressures.

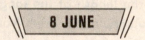

8 JUNE

Why did American cowboys shoe their horses, but Native Americans did not?

THE HORSE'S NATIVE habitat is large grassy plains with a generally dry climate. Horses were driven to extinction in North America about 7,600 years ago, possibly by climate change or hunting by the ancestors of Native Americans. They only returned to the New World when the Spanish brought them there in the 16th century.

The tribes of the North American plains and the American Southwest came across these horses, or at least their feral descendants, after they began to escape from Europeans around 1540. Newly established wild herds spread up the Mississippi valley, where most of the tribes had a settled agrarian lifestyle. The Plains Indians led a nomadic existence, and despite never having seen a European – mounted on a horse or otherwise – it was they who realised the horse's potential for enhanced mobility. In the space of just over a century the horse had transformed Native North American society.

The horseshoe was developed to meet the conditions faced by domestic horses in north-west Europe, where, judging from archaeological evidence, they were probably first produced by the Gauls or Franks in the 5th century. Europeans needed horseshoes because of a combination of climate, terrain and pattern of use, with the generally wet weather and soft, heavy soils acting to soften the normally calloused sole of the hoof. Horses were used for travel and in wars. They were often heavily laden while travelling at quite high speeds, which placed great stress on the hooves, often causing them to wear unevenly and eventually split, rendering the animal lame and useless.

The lifestyles of horses used by Plains Indians, on the other hand, differed little from that of their ancestors in the wild. The animals moved together in large numbers at relatively low speeds, over flat, arid steppe country. As a result, their hooves were harder and wore more evenly. In addition, Native American warriors had more than one mount each, with one band of 2,000 Comanche braves keeping a string of 15,000 horses in tow. A Plains Indian brave's horses were his fortune and livelihood, and he cared for them accordingly if he valued his life.

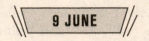

What happens when you lose consciousness?

CONSCIOUSNESS FEELS LIKE an on–off phenomenon: either you're experiencing the world or you're not. But finding the switch that allows our brains to move between these states is tricky.

One common definition of consciousness is 'the thing that abandons us when we fall into a dreamless sleep, and returns when we wake up'. But say someone anaesthetises you: you may hear their voice, but not respond to it; you may be dreaming and not hear their voice; or you may hear or experience nothing at all. What patterns of brain activity correlate with these levels of conscious experience?

We do know there are certain brain regions that, if damaged or stimulated, cause loss of consciousness. The claustrum – a thin, sheet-like structure buried deep inside the brain – is one of them. But many leading theories that aim to describe consciousness veer away from a single anatomical site being its seat.

Global workspace theory hinges on the idea that information coming in from the outside world competes for attention. We only become conscious of something – a ringing telephone, say – if it out-competes all else to be broadcast across the brain.

Then there's information integration theory, which suggests that consciousness is the result of data being combined to be more than the sum of its parts. A recent study that scanned people's brain activity as they were slowly anaesthetised seems to back up that picture. It might also explain how a drug like ketamine knocks people out: this potent tranquilliser ramps up activity in many of the brain areas that promote wakefulness, but depresses communication between different regions.

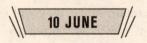

How high can a butterfly fly?

UNLIKE HUMANS, BUTTERFLIES are not disposed to seeking altitude records. Indeed, they will not fly higher than is strictly necessary in their everyday lives, whether looking for a mate, food or somewhere to lay eggs, avoiding predators or migrating.

Worldwide there are many thousands of species of butterfly, each adapted to its own particular habitat and needs. Some spend their whole lives on a patch of coastal grassland, the larvae feeding on low plants or living in ants' nests, and the adults never flying more than a few feet above the ground. Others are only found on high mountains. Even though they don't actually fly very high above the ground, butterflies that live on the mountains of Peru spend their whole lives at altitudes of around 6,000 metres.

Butterflies that migrate tend to fly the highest in general. The most famous migratory butterfly is probably the monarch, known to scientists as *Danaus plexippus*. These leave Mexico or California each year and fly north to Canada or the northern US, though actually it takes several generations to get there. Monarchs have been sighted by glider pilots flying as high as 1,200 metres. Interestingly, they seem to fly in the same way as a glider, using updraughts to gain sufficient altitude so that they can glide for quite a distance before needing to use energy to climb again.

Europe also has plenty of migratory species. The painted lady, *Vanessa cardui*, makes its way to southern France from north Africa; it has to leave Europe in winter because it can't survive a frost. To get to France many will cross through the mountain passes of the Pyrenees, which in general lie at about 2,500 metres. If they encounter a high building, they just fly straight upwards and over it. If they encounter a high mountain range, they will do the same.

However, insects of any kind cannot fly if they are too cold. Butterflies can keep warm by beating their wings, though if they fly too high in the wrong conditions, they may become chilled. On average, the air temperature reaches freezing at an altitude of just below 8,000 metres, suggesting that this would be their physical altitude limit.

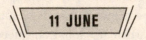

11 JUNE

Why do biscuits go soft, but bread goes hard?

LEAVE A BISCUIT out overnight and it will become soft by the morning; but leave a baguette out for the same length of time, and it will become so hard you could knock someone out with it. What causes these opposite reactions?

Biscuits contain much more sugar and salt than baguettes. The finely divided sugar and salt are hygroscopic and soak up moisture from the atmosphere – the osmotic pressure in a sweet biscuit is quite high. The dense texture of a biscuit helps maintain the moisture by capillary effects. Try a series of different biscuits, varying from very sweet, dense ones to light, fluffy sponge biscuits. You'll find that the 'overnight sogginess index' increases with density and sugar/salt content.

The baguette, on the other hand, contains little salt or sugar, and has a very open structure. Without sugar, there is nothing to attract water vapour, which evaporates to leave the baguette hard. The process is temperature dependent, with the rate fastest at just above freezing and slow below freezing. Studies show that bread stored at 7 °C (average fridge temperature) becomes stale at the same rate as bread stored at 30 °C. So putting bread in the fridge does not keep it fresher for longer.

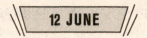

12 JUNE

Why do animals lick their wounds?

LICKING EACH OTHER'S and their own wounds is the most common form of wound treatment for mammals. It is believed that such behaviour dates from the earliest days of mammals. Saliva generally is germicidal and benefits wound tissue, causing little harm to live tissue while helping to slough off or recycle dead tissue.

The habit no doubt developed out of a defensive response to the pain, plus an eating response to bodily fluids and detritus. In fact, when mothers of many species lick sick cubs, if there is no improvement, it can lead to them eating their babies. Distressingly, such disruption may also lead the mother to eat the rest of the litter, especially if they are very young.

Formal hygiene and treatment of illness and injury, especially of other individuals, is mainly a human behaviour. However, it depends on what you choose to call 'treatment'. Candidate activities among birds include dust-bathing, hiding and resting when ill, and 'anting' – where they rub their feathers with ants, which then secrete antimicrobial chemicals. Various birds and mammals eat clays to counteract poisons in food, and some types of chimpanzee chew certain pungent leaves when ill. Such 'medicines' may control parasitic worms. Since plants and traditions vary by region, those habits clearly get passed on as learned knowledge.

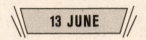

Will we all speak the same language one day?

THE STEADY ADVANCE of technology and globalisation mean that a few languages dominate and grow, while many others decline.

With over a billion native speakers, Mandarin Chinese is the language spoken by the greatest number of people. English comes third, after Spanish. But unlike Mandarin and Spanish – both spoken in more than thirty countries – English is found in at least a hundred. In addition to the 335 million people who speak it as a first language, 550 million cite it as their second. It dominates international relations, business and science. All this suggests that English is on course to become the planet's lingua franca. It just probably won't be the English that native speakers are used to.

Millions of second-language English speakers around the world have created dialects that incorporate elements of their native languages and cultures. These varieties are known as similects: Chinese-English, Brazilian-English, Nigerian-English. Taken together, they – not American or British English – will chart the language's future path.

Even in a future where China, India and Nigeria are global superpowers, English is likely to be the language of choice for international discourse, simply because it is already installed. Weirdly, this is not such good news for native English speakers. When all the people of the world have English, it's no longer a special thing, and native speakers lose their advantage.

In time, English similects may begin to blend over national borders. New dialects are likely to form around trades or regions. These common goals will drive the evolution of the lingua franca, regardless of whether we call it English or not. That is not to say that all other languages will vanish. German will remain the language of choice within German borders. Even Estonian, spoken by just 1 million people, is safe. Likewise, the language directly descended from Shakespeare's English has staying power with Brits and Americans. But English, like football, will soon move outside their control, pulled into something new by the rest of the planet.

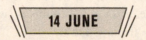

14 JUNE

Is eating bogeys bad for you?

PICKING AND EATING are for cherries. Even if eating your bogeys isn't bad for you, chewing them could affect the health of your friendships! Try chewing gum instead.

Physiologically, eating your dried snot would not matter much. If that solid bogey that you find so toothsome had not dried out, it would have dribbled down your pharynx and been swallowed anyway, unless you intercepted it with your sleeve or handkerchief or stuck it on the underside of your chair.

For the most part, any germs it contained would be digestible or would otherwise die in your gut, but this is not always the case. Some germs do infect people via the nose, and some toxic dust particles do stick in your phlegm. It is to the benefit of your health to ensure you expel such things.

It is not for nothing that your nose hairs stop bugs and dust from landing in your lungs or gut. Blowing your nose would not stop everything, but it is better than guzzling snot.

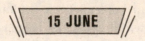
What happens when an incandescent lightbulb blows?

INCANDESCENT LIGHTBULBS USUALLY blow when you turn them on, rather than at the end of a long evening's use, which seems paradoxical. But when one of these lightbulbs is switched on, its delicate filament is hit with a triple whammy.

The resistance of the metal filament rises with its temperature. When it is switched on, its resistance is less than a tenth of its usual working level so an initial current more than 10 times the rated value surges into the filament, heating it very rapidly and creating thermal stress.

If any part of that filament is slightly thinner than the rest, this area will heat up more quickly. Its resistance per millimetre will be higher than that of the rest of the filament, so more heat will be generated along this stretch than in the adjacent parts of the filament, amplifying the thermal stress.

In addition to all this, the filament is wound into a coil which also acts as an electromagnet. Because of this magnetism, each turn of the coil deflects its neighbouring turns so that the initial surge of current jolts the thin and delicate filament, creating mechanical stress.

So it is no wonder the poor thing goes splat when you throw the light switch.

Can I move a ship with my bare hands?

IF THAT SHIP is floating freely, with no wind or current, then it is in fact remarkably easy for a person to move a large ship, such as the *QE2*. This can be explained in terms of kinetic energy (E) and momentum.

Consider a ship with a mass of 20,000 tonnes ($m = 2 \times 10^7$ kilograms). If the ship is given a velocity of 1 centimetre per second ($v = 10^{-2}$ metres per second) then its energy, $E = 1/2\ mv^2 = 1/2 \times 2 \times 10^7 \times (10^{-2})^2 = 1,000$ joules. A thousand joules is a very modest amount of energy. It is the energy expended by a 51-kilogram man climbing up a 2-metre-high flight of stairs.

At 1 centimetre per second the ship's momentum (mass × velocity) $= 2 \times 10^7 \times 10^{-2} = 2 \times 10^5$ newton-seconds. The 51-kilogram man can impart this to the ship by applying his full weight for 400 seconds: $51 \times g \times 400 = 2 \times 10^5$ newton-seconds, where g is the acceleration due to gravity, 9.8 ms^{-2}. If he moves the ship by standing with full weight on one of the mooring lines, he will have descended by 2 metres by the time the ship is moving at 1 centimetre per second.

Actually when a ship is set in motion, a comparable mass of water is also set in motion at a comparable speed. Consequently the kinetic energy and momentum calculated above have been underestimated by a factor of two or so. However, the main conclusion stands: an unaided person can easily move a ship.

So push, and good luck!

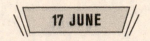

What causes hay fever?

THE RESPONSES THAT cause the symptoms of hay fever are nearly identical to the body's mechanism for destroying and expelling parasites. How these reactions come to be triggered by innocuous substances like pollen is not well understood. What is known is that the sensitisation process generally starts months or years before you notice any symptoms. The end result is a network of immune cells primed to react to specific allergens when you encounter them again in the future.

For some reason, more and more people are being sensitised. Allergic rhinitis affects people across the world but the highest recorded incidence among adults is 30 per cent in the UK, three times what it was in the 1970s. The US, Australia, New Zealand and other Western countries have experienced similar growth. The global average is about 16 per cent, meaning that more than a billion people have hay fever.

Symptoms can suddenly start at any time of life, even in old age. Drinking alcohol and smoking also increase the risk of developing hay fever and make symptoms worse. Increased awareness could be behind this epidemic, but that's not the whole story – allergies of all kinds have seemingly become more common in recent decades.

The leading explanation is the hygiene hypothesis – the idea that decreased exposure to bacterial infections and parasites during early childhood disrupts the normal development of the immune system, causing it to pick fights with harmless substances. Growing up on a farm seems to protect children against hay fever and asthma, as does drinking unpasteurised milk.

But the hygiene hypothesis is not a complete explanation either.

Certain countries such as Japan are extremely Westernised yet have very low rates of allergy – possibly because of underlying genetic factors that influence susceptibility. If you have one close family member with hay fever, you have a greater than 50 per cent chance of developing it yourself.

One recent twist is the suggestion that reduced contact with natural environments might lower the diversity of microorganisms living in and on us, and that this might tip the immune system towards an allergy-prone state. Ilkka Hanski at the University of Helsinki in Finland recently discovered that, compared with healthy individuals, people who are predisposed to developing allergies are more likely to live in built-up areas, and have less-diverse bacteria living on their skin.

One group of bacteria called *Acinetobacter* looks especially interesting because it seems to encourage immune cells to produce an anti-inflammatory substance called IL-10. Children growing up in homes that are surrounded by more forest and agricultural land tend to have more of these bacteria on their skin, and lower rates of allergy. Other possible factors that increase susceptibility include greater use of antibiotics during childhood, lower levels of vitamin D or exposure to certain chemicals.

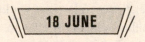

18 JUNE

Does every tree have a unique rustle?

ANY AIRFLOW DISTURBANCE, such as that caused by leaves, creates sounds of characteristic volume, frequency and oscillation. Trees' songs change with wind speed and direction, and the type of leaves.

Needle-like leaves shed vortices as the wind oscillates round them, creating the high-pitched, romantic whisper of conifers. Flat leaves flap like flags, depending on thickness, firmness, edge outline and surface texture. This is commonly the main component of the rustling sound. Pointed, narrow willow leaves shed wind energy with whisperings.

Colliding leaves suffer damage, so they grow in patterns to avoid touch. In high winds, though, impact is inevitable, causing another kind of rustling. Smooth, large, simple leaves tend to give low notes except when flapping vigorously. Trees with small leaves, prominent veins, complex outlines, furry surfaces and rough bark seem quieter, but produce ultrasonic sounds.

Crisp autumn leaves act as rattles. Hollow leaves emptied by aphids, and acacia thorns hollowed by ants, may whistle. Dense foliage dampens high notes. Leaves on high branches differ in shape and texture, and encounter higher winds. The leaves of rushes scrape and vibrate like the reeds of wind instruments, giving rise to the Greek legend about their whispering: 'King Midas has ass's ears!'

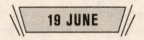

19 JUNE

What would happen if we found aliens?

THANKS TO THE Kepler Space Telescope, we know the galaxy could hold as many as 30 billion planets similar to our own. Some think it's just a matter of time before we find out we're not alone.

What we detect will make a big difference to how we react. Any discovery that is less obvious than little green men landing during the World Cup final is likely to be met by years of questions and examination, so there may not be an 'aha' moment.

A chemical imbalance in an exoplanet's atmosphere could be a sign of microbial activity. But an indirect result such as this will probably have only a short-term impact. The apparent discovery of

Martian nanofossils in meteorite ALH84001 in 1996 led to a media frenzy, and even US congressional hearings, before the furore died down in the face of increasing scepticism. Most now think that the meteorite does not hold the remnants of ancient alien life.

A decoded broadcast from intelligent aliens would be altogether different. Scientists and governments would have to assess whether the message was threatening and what, if anything, should be sent as a response. It would pose a challenge to certain religions too. Some people might see intelligent aliens as saviours themselves, giving rise to new religions. Others may simply celebrate them as species that overcame their provincial squabbles to explore the universe.

In the longer term, even slight evidence of extraterrestrial life would spark a quest to understand the universal principles of biology. We may find answers to questions such as: does life arise wherever the conditions are right, or is it a freak accident? Are there other types of genetic code? Does life always require carbon or water? Is Darwinian natural selection a universal, or are there other forms of evolution?

Perhaps most significantly, it would be the decisive blow to the idea that humans are the centre of the cosmos or the reason for its existence. Instead, we would be forced to acknowledge our place as just one tiny branch of a vast galactic tree of life.

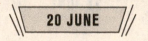

20 JUNE

In running races, how is the winner decided in a photo finish?

IN RUNNING RACES, the winner is the athlete whose torso crosses the line first. Hands, knees and heads don't count.

The 'beam' of the finishing line is really a camera that rapidly and repeatedly takes a picture of the plane of the finishing line. Over time, the slices of images of the finishing line are compiled into one picture and coordinated with the event clock to recreate a time-lapsed picture of the finishing line.

These pictures are how the times and places of the athletes are determined. They can be odd in appearance. For example, if an athlete's foot lands on the finish line itself, multiple photo slices give it the appearance of a ski.

While the picture itself is auto-generated, the times and places are determined by a human judge. For each athlete the judge finds the position of the athlete's chest. While this might not sound like a well-defined point, the athletes know how to contort their bodies at the last instant to make the chest cross the finish line at the earliest possible moment. Once the judge pinpoints the athlete's chest the corresponding time of crossing the finishing line is determined from the picture.

To minimise athletes blocking each other, the camera is raised and gets a bird's-eye view of the line.

There would probably be far more injuries from diving at the line if the hand rather than the chest were chosen to determine the winner.

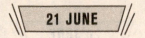

21 JUNE

Why are balloons harder to inflate when fully deflated?

SOME PEOPLE HAVE the knack of blowing up balloons, while others fail to get any air into them at all. Well, those in the latter group, there may be a scientific justification for your struggle.

Balloons can teach us quite a bit about physics. Try this: blow up two identical balloons to different sizes and pinch the ends so that the air does not escape. Connect their openings with a piece of plastic or rubber tube using sticky tape and the help of a friend if it proves too difficult on your own. Release the pinched ends so air can flow freely through the tube.

Intuition tells you that the balloons will equalise in size. But they don't. Instead, the small balloon shrinks as it forces its air into the bigger one. A number of theories have been propounded to explain why this happens, and it is likely that all contain some truth. The common theme is that the air moves from the smaller balloon to the larger balloon because the pressure inside a balloon increases as its radius decreases. For a compelling demonstration, stretch the opening of an inflated balloon over a party squeaker. As the balloon deflates, the pitch emanating from the squeaker rises rather than falls, because the air is forced out ever faster. This suggests the pressure in the balloon does indeed rise as it shrinks.

Assuming we accept this, the question is: Why? Some believe it's because balloons act like bubbles, which also display this relationship between pressure and radius. However, balloon rubber does not behave in the same way as a soap film. The tension in the rubber changes in a non-linear fashion – that is, it is not a simple function of how much the rubber has been extended, and common experience supports this. The most obvious example is when you start blowing up a balloon. The initial extension of the rubber requires a lot of puff. Once you do, however, the balloon inflates quite easily. This indicates that the tension in the rubber exerts a higher pressure at the initial low-extension stage than it does when it is much larger, which would account for the smaller balloon expelling its air into the larger balloon.

So if you struggle to blow up a balloon, feel vindicated.

Does breakfast cereal really contain iron?

BREAKFAST CEREALS OFTEN claim to be fortified with iron – and if you have a magnet, you can extract it too! So ponder the ingredients list on your packet of cornflakes while you are munching breakfast, and then set about removing one of them by following these steps.

First, fill a plastic cup to about two-thirds full with iron-fortified cereal, and with a spoon or a pestle crush the cereal into a fine powder. It is worth spending a lot of time on this stage – the finer the powder, the better. Then, put the crushed cereal into a clear, sealable sandwich bag and add hot water. Leave the mixture for about 15 to 20 minutes.

Now gently tilt the bag forward so that the cereal collects on one side, and place a magnet along the outside of the bag near the cereal, running it over the bottom, because the iron tends to sink. Tilt the bag back so that the cereal runs away from the magnet. You can also lay the bag flat on the table and stroke it with the magnet towards one corner. The magnet will attract a black fuzz of iron. Move the magnet over the surface of the bag or blender and the tiny pieces of iron will follow it.

The black stuff in your cereal really is iron – the same stuff that is found in nails and trains and motorbikes. And it's quite heavy, which is why you need to make sure you run your magnet along the bottom of the cup. The iron is added to the mix when the cereals are being made and you really do eat it when you devour your cornflakes.

The reason it is added in a form that you can extract is that iron ions (iron that would more easily combine with other molecules in the cereal) would increase the spoilage rate of the food. Using iron in its pure metal form gives the cereal a longer shelf-life. The hydrochloric acid and other chemicals in your stomach dissolve some of this iron and it is absorbed through your digestive tract, although much of it remains untouched and comes out in the loo.

Humans need iron for many important bodily functions. Red blood cells carry haemoglobin, of which iron is a key constituent. Haemoglobin transports oxygen through the blood to the rest of the body by binding oxygen to its iron atom and carrying it through the bloodstream. As our red blood cells are being replaced constantly, iron is an essential part of our diet.

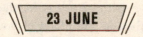

23 JUNE

Why do some birds stand on one leg?

SOME PEOPLE THINK that the reason flamingos stand on one leg is so ducks don't swim into them as often! Nice idea, but the most likely answer has to do with conserving energy.

In cold weather, birds can lose a lot of heat through their legs because the blood vessels there are close to the surface. To reduce this, many species have a counter-current system of intertwined blood vessels so that blood from the body warms the cooler blood returning from the feet. Keeping one leg tucked inside their feathers and close to the warm body is another strategy to reduce heat loss.

The converse is probably true in hot climates – blood in the legs will heat up quickly, so keeping one leg close to the body will reduce this effect and help the birds to maintain a stable body temperature.

Another factor in long-legged birds is that it may require significant work to pump blood back up the leg through narrow capillaries. Keeping the leg at a level closer to the heart may reduce this workload.

It is also worth remembering that birds' legs are articulated differently to ours; what looks like the knee is in fact more like our ankle. Many birds have a mechanism to 'lock' the leg straight, so for them it is much easier to stand for hours on end on just one leg – and even take off from and land on one leg.

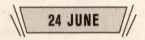

24 JUNE

Could blood transfusions reverse ageing?

IN THE 16TH century, so the story goes, the Hungarian countess Elizabeth Báthory bathed in the blood of young girls in a bid for eternal youth. Without condoning Báthory's deplorable sadism in any way, it is a curious question: if one were to take a blood sample from an infant, store it perfectly for 50 years, then reintroduce it to the body of the adult, could it have any positive effect?

To answer this, we first need to ask ourselves: what are the causes of ageing? Ageing is a complex interplay of many different phenomena including a gradual decrease in mitochondrial function because of oxidative stress and the build-up of misshapen proteins resulting from transcription mistakes and accumulated DNA damage.

The second question is whether a transfusion would work. Infants can breathe a sigh of relief, because the answer is no. Replenishing 'aged' blood with 'young' blood would not ameliorate any of the cellular phenomena that lead to ageing. The most probable outcome would be negative: the person involved would quickly become sick after the transfusion because the replaced blood would lack the circulating antibodies that the individual had built up over the preceding 50 years. As a result, germs that had not been a problem prior to the transfusion would suddenly find a new and easy target in the new blood circulating through the body.

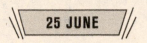
Why can't we see the footprints on the moon with a telescope?

THE RESOLVING POWER of a telescope – the size of the smallest object it can see at a given distance – is inversely proportional to the diameter of its lens. In other words, to see something small a long way off you need a very big telescope.

We can see distant galaxies but cannot see the much closer footprints left on the moon because galaxies and galaxy clusters represent a bigger target: they take up, or 'subtend', a much larger angle in the sky. Galaxies are also bright, making them stand out against the blackness of space. Footprints are simply impressions left on the lunar surface, offering no contrast at all. We would be reduced to looking for shadows cast by the tread.

Apollo 11's Eagle lunar module measures about 4.3 metres across, and to see it from Earth, when we are at our closest to the moon, would require a telescope with an angular resolution of 670 billionths of a degree. If we take the wavelength of the reflected light from the moon as being 550 nanometres, the middle of the visible range, then to see the lunar module would require a telescope with a diameter of nearly 60 metres. The largest telescope now in existence, the Gran Telescopio Canarias on the Spanish island of La Palma, has a diameter of 10.4 metres.

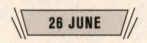

26 JUNE

How much mucus do you produce when you have a cold?

WHEN YOU HAVE a cold, it can seem like you need to blow your nose a million times. But how much do you actually produce?

On average the normal nose produces 240 millilitres (about a cupful) of mucus every day. During a cold there is additional flow. Most of the mucus produced normally flows down the throat, and gets recycled within the body. During infection the nasal passage becomes constricted and therefore the inward passage of mucus is obstructed and it flows out through the nostrils. Over the course of a day a mild cold might produce a few millilitres of snot for blowing. Heavier colds, demanding the handkerchief 20 times an hour at say 2 to 10 millilitres per blow, could cost you as much as 200 millilitres an hour, and you then must drink liquids to make up the volume.

There are other factors that increase the mucus flow, like the excessive formation of tears, which can make their way to the nasal passage and mix with the mucus. Nasal flow usually slows at night, but really serious secretion can force one to sleep sitting up to avoid swallowing phlegm and saliva.

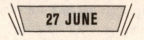

27 JUNE

Can bees always find their way home?

IN MOST CASES, yes. Bees employ special orientation flights to memorise near and distant landmarks relative to the nest. As well as these cues, they also use the sun's position as a landmark, making use of an inbuilt clock to compensate for the sun's movement across the sky.

In one study, tagged worker Bombus terrestris and Bombus pratorum bees were released at staged points from their nests and all returned safely, the farthest from 6 kilometres away. The bees were taken to their starting points by car, in the expectation that their usual navigational aids would be compromised, much as one might expect from a train journey. However, it is likely that their internal clock, compensating for the movement of the sun, enabled them to use solar positioning to fly back to a distance from the nest at which their visual cues would come back into play.

The ability to find a nest from a long distance away is vital to bees because nest sites and food may not be found in the same habitat. It is also known that workers of the much smaller honeybee Apis mellifera can forage up to 13 kilometres from the hive, even in wooded country, and females of some large neotropical bees are thought to have foraging ranges of up to 30 kilometres.

If a queen Bombus bee gets lost, it might be able to insinuate itself into an established colony of the same species by lying low in the nest to give it time to absorb the colony odour and avoid aggressive responses from resident bees. An egg-bound queen with aggressive tendencies might kill the resident queen and assume control of the colony. But an exhausted and disoriented bee entering a strange colony may well be killed by the workers. All of these behaviours have been reported for bumblebees.

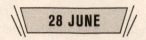

What happens when a black hole is swallowed by another black hole?

WHILE THIS QUESTION sounds like it might reach the far reaches of the scientific imagination, it is not quite as exciting as you might think.

If a black hole were to be swallowed by another, the two black holes would fuse, creating a new one with a mass equal to the two constituent masses. The surface area of a black hole is related to its mass so this would increase in proportion. Because black holes are generally accompanied by an accretion disc of material that rotates around them moving faster and faster as it spirals inwards we can assume that the two discs will crash into one another before the holes meet. The end result will be a bigger disc but when the two very hot discs hit each other there will be an increase in the radiation given off until the new system settles down.

When two black holes approach one another they usually follow hyperbolic paths around a common focal point so they end up galloping off in different directions. If there are stars around, the situation becomes more complicated, and the black holes can end up orbiting one another after tossing some stars away. Such an orbit will continue almost forever unless there are still more stars nearby that can be thrown off, removing some energy and allowing the black holes to fall closer together. A black hole may also head straight towards another so that it would hit its event horizon. In that case, an observer would see the two black holes getting closer and closer (assuming they're visible), but would never see them coalesce. Instead, they would seem to slow down, get dimmer, and the light from them would get redder. This is because their time slows down so we might never see them reach the point at which one crosses the other's event horizon, even though that happens in due course.

Since 2015, we've been even more sure this scenario isn't science fiction. That was when scientists at the Laser Interferometer Gravitational Wave Observatory, or LIGO, in the US first detected ripples in space-time that Einstein's general theory of relativity predicts should be sent out when very massive objects move around in the universe. The size and shape of the signal was just what scientists expected from two black holes merging – and there have been lots more similar detections made since.

29 JUNE

Why are orangutans orange?

IN THE FOREST being green helps animals to go unnoticed. Orangutans live in the rainforests of Borneo and Sumatra – which are also rather green – so why are they orange? Surely that makes them easy to spot?

But orangutans' colouring helps them blend in. The water in the peat-swamp forests where orangutans live tends to be a muddy orange. Sunlight reflected off this water can give the forest an orange cast, making orangutans surprisingly hard to see in dappled light. Many orangutan nests, up in the forest canopy, contain orangey-brown dead leaves, and some trees have reddish leaves, especially when young.

Ground-based predators would see orangutans in the canopy as a mere silhouette. In such circumstances, orange may stand out less than black, which may be more suited to blending in with the forest floor.

Dark African apes such as gorillas spend much more time on the ground than orangutans, while some other canopy-dwelling primates have a similar ruddy colour to orangutans. Among these are red langurs, which live in the same Borneo forests as orangutans.

30 JUNE

Why do objects close up move faster than those far away?

YOU CAN TRY this one out when you're on the train or in a car on the motorway.

First, you'll see that objects further away look smaller. You can use your hands to show this: if you hold one hand close to your face and the other at arm's length, the one at arm's length will appear small, even though they are (probably) the same size.

Second, you'll notice that it takes more objects to fill the same amount of visual space if those objects are further away. For example, if the hand further away is half the apparent width of the one closer, it takes two hands to fill the same width.

Finally, think about something moving, such as your index finger traced slowly from one side of your palm to the other. If it moved at the same speed when it was further away, it travelled the same actual distance (a palm's width), but seemed to have travelled only half as far. So it would take twice as long for it to look like it had travelled the same distance. Distant things are not actually moving slower, they just look as if they are.

The reason why is because our field of vision is shaped like a cone, with the small end at our eye and the big end at the very limit of what you can see. This type of optical system causes us to see a particular object as 'smaller' the further away it is. This is called foreshortening.

JULY

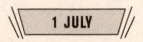

Why do chillies burn for longer than mustard?

MUSTARD AND CHILLIES are both hot, but the burning sensation from a chilli stays in the mouth for ages while the sensation from hot mustard disappears in a few seconds.

This is because the chemical mainly responsible for the burning spice in chilli peppers is capsaicin, a complicated organic compound that binds to receptors in your mouth and throat, producing the desired (or dreaded) sensation. Capsaicin is an oil, almost completely insoluble in water. This is why you need a fat-containing substance like milk to wash it away – watery saliva doesn't do the trick.

On the other hand, the compound responsible for the hotness of mustard (as well as horseradish and wasabi) is called allyl isothiocyanate. This chemical is slightly water-soluble, and can be more readily washed away into the stomach by saliva.

Further, the chemical in mustard is more volatile than capsaicin so it evaporates more readily, allowing its fumes to enter the nasal passages (explaining why the burning sensation from mustard is often felt in the nose). These fumes can be easily removed by breathing deeply, a useful strategy if the sensation becomes overwhelming.

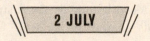
Do polar bears get lonely?

HAVING A GREGARIOUS or solitary nature are species-specific survival strategies adopted by different animals and birds. Big predatory mammals such as polar bears, grizzlies and tigers isolate themselves from one another to avoid competition with other members of their own species. By spreading out, they also expand their feeding grounds and breeding territories. If fellow species members come into close proximity there can be fierce competition for food, mates and territory. The same is true with many solitary species of bird, such as eagles and condors.

These animals and birds usually pair up during the breeding season to reproduce, and separate soon after successful mating or when they have raised their young ones. In most cases, raising the young is the sole responsibility of females. Indeed, males of such species sometimes kill their young to increase their own reproductive success.

Social animals, by contrast, find strength in numbers. Animals such as antelope on the African savannah or penguins in the Antarctic form big colonies, where they huddle together for warmth and to alert each other to a potential predator attack. In a large herd or colony, losses to predators are negligible compared with what they would be if the animals were in isolated groups.

Between the solitary and social extremes are creatures like lions, wild dogs and wolves, which often hunt in groups and display differing degrees of social interaction and cooperation.

A similar question can be asked about why some plants are gregarious while others are solitary. In one intriguing strategy, called allelopathy, gregarious plants secrete chemicals into the soil to reduce competition from related species that cannot survive the presence of these compounds. As with animals, these strategies have evolved to maximise the plants' chances of survival.

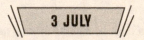

3 JULY

Do our fingerprints change?

EXPERTS REFER TO the patterns that make up fingerprints as friction ridges. These begin to form in developing fetuses roughly halfway through pregnancy, and differ even between identical twins, and conditions in the uterus contribute to the pattern. With few exceptions, these patterns are permanent.

Some people are born without fingerprints. As of 2011 members of five families are known to have adermatoglyphia, caused by improper expression of the SMARCAD1 protein. People with Naegeli–Franceschetti–Jadassohn syndrome and dermatopathia pigmentosa reticularis, which are both forms of the condition ectodermal dysplasia, also lack fingerprints.

Fingerprints can be temporarily erased by physical abrasion or medication, and because skin elasticity decreases with age, they can be difficult to record in older people. There are also criminals who try to evade detection by burning off their fingerprints or smoothing them out using glue and nail varnish. Notorious 1930s US gangster John Dillinger used acid in an attempt to erase his, and another, Robert Phillips, grafted skin from his chest onto his fingers to do the same – but was convicted using prints from other parts of his hands.

It has recently come to light that fingerprint analysis, once regarded as the gold standard in forensic science, is fallible. In his book *Finger Prints*, Francis Galton, cousin to Charles Darwin, calculated that the chance of a 'false positive' (two individuals having the same finger-prints) was about 1 in 64 billion. This may be true, but does not take into account that such analysis is not a perfect science, and relies on judgement, which is susceptible to cognitive bias. Fingerprint analysis is a twist on the spot-the-difference picture games we play as children, except here you are looking to match up as many as fifty 'landmarks' (for example where a ridge splits in two), often with a smudged or incomplete print from a crime scene.

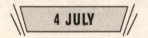

4 JULY

How do astronauts navigate in space?

NAVIGATION REQUIRES THAT you know where you are relative to your destination, and how to negotiate the route. On Earth, if you want to get from point A to point B, you can follow a compass to get there. To do this in space, knowing your attitude is as crucial as knowing your position, so the first thing to do is find and track the sun and a known, conspicuous, distant star.

Sirius is a good example, but it lies relatively near the celestial equator, so sometimes the sun gets in the way. Canopus is better: almost as bright, and far south in the celestial sphere, well away from the sun. From the positions of such stars and the sun, you can calculate your attitude, and can locate other bodies by radar, data from mission control or visual observation. Gyroscopes can then damp oscillations and detect minute changes in attitude, while Doppler measurements let you calculate your velocity.

In space, knowing your trajectory relative to major masses in the solar system permits you the luxury of navigating by dead reckoning for millions of kilometres. Only when you apply thrust to adjust your course, get close enough to bump into things, or need to manoeuvre into a precise orbit is it necessary to check on your exact situation and make corrections.

The Apollo missions depended on ground-based radar, which could determine their position and range, plus, using Doppler measurements, their radial velocity. Course changes were computed on the ground and radioed to the crew. The figures were then punched into the on-board computer, which took care of the engine burns. As a back-up, the crews were trained to be able to take star readings and calculate course changes for themselves. This was never required, although for one course adjustment on Apollo 13, the computer was not available and the burn had to be controlled manually.

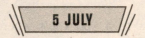

5 JULY

Why does hot water freeze faster than cold water?

THOUGH IT MAY seem to defy common sense, astoundingly this is true – and you can see it for yourself at home.

Fill one flattish plastic container or tray with the warm water (about 35 °C), a second with the cool water (about 5 °C) and place them in the freezer. Give them 10 minutes at least before checking them, then check again at regular intervals. The warm water will freeze faster than the cold water.

This seems paradoxical, but can be explained by understanding that there is a vertical temperature gradient in the water. The rate of heat loss from the upper surface is proportional to the temperature of the surface. So, if the surface can be kept at a higher temperature than the bulk of the remaining liquid (in this case, the initially warmer water), then the rate of heat loss will be greater than from liquid with the same average temperature uniformly distributed (in this case the initially cooler water). This can be proved by using tall, thin metal cans as water containers. The paradoxical effect disappears in this case because the temperature gradients in the cans are lost as heat is rapidly conducted through the metal sides much faster.

Nonetheless, there may be other factors at work. Cold water forms its first ice as a floating skin which impedes further convective heat transfer to the surface. Hot water, on the other hand, forms ice over the sides and bottom of the container, while the surface remains liquid and relatively hot, allowing radiant heat loss to continue at a faster rate. The large temperature difference drives a vigorous convective circulation which continues to pump heat to the surface even after most of the water has become frozen.

There could be yet another factor affecting this process – super-cooling. Research shows that because water may freeze at a variety of temperatures, hot water may begin freezing before it is cold. But whether it will completely freeze first may be a different matter.

It seems that a number of factors are at work here, with no one theory acting in isolation. Do take time to try out any number of combinations of temperature, vessels and environments – fame awaits whoever stumbles on the ultimate answer.

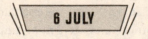

6 JULY

How do relighting candles work?

FOR A FLAME to exist, it needs oxygen, fuel and enough heat to keep combustion going once it starts. These requirements are sometimes represented graphically as the three sides of a 'fire triangle'.

When you blow out an ordinary candle, you extinguish the flame by removing heat. You can still see the fuel – the wax smoke, often paraffin vapour – coming from the wick, but the ember in the wick does not supply enough heat to reignite it, and so it will eventually go out.

But when the wick has magnesium powder in it, as is the case with relighting candles, the ember is able to ignite the magnesium. This element catches fire at relatively low temperatures, below 500 °C, and this is enough to reignite the smoky fuel. The flame itself then burns at around 3000 °C.

If you look closely at one of these candles after you have apparently blown it out, you can see the smouldering wick emitting little sparks of burning magnesium before the candle relights.

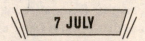

7 JULY

Why don't you see the blood vessels in your eyes?

MAKE A PINHOLE in a piece of cardboard. Bring your eye close to it and look through the pinhole as you rotate the card. You will see the network of your retinal capillaries against the background of a cloudy sky.

This is a fascinating phenomenon known as Purkinje shadows, after the Czech physiologist and neuroanatomist Jan Evangelista Purkinje. It also illustrates an excellent argument against intelligent design.

In the human eye, light passes through all the nerve fibres and blood vessels before reaching the photoreceptors. This curious arrangement means that the blood vessels cast shadows on the back of your eye, and it explains why the capillaries can be seen if you look through a moving pinhole. Surely a master creator wouldn't have made a mistake like that. After all, the squid eye is designed the other way around, which raises the possibility that the mythical

210

intelligent designer of life considers cephalopods a higher form of life than humans.

The reason we don't normally see the shadows of the blood vessels is because the human eye is incapable of registering a stationary image. We can see things that don't move, such as statues or doors, only because our eyes are continually making tiny movements which ensure that their image jiggles across our retina. Using sophisticated eye-tracking equipment it is possible to completely stabilise any retinal image. When this happens the image disappears, a phenomenon called Troxler's fading. If our eyes were completely still we'd be almost blind. Intelligent design, huh? Because the blood vessels are part of the eye, they move with it – meaning they are essentially stationary as far as our photoreceptors are concerned – and therefore usually remain invisible.

Moving a pinhole across the pupil changes the direction of light reaching the back of the eye, which has the effect of 'moving' the capillaries relative to the retina, making them visible. An even better way to see the blood vessels in your own eye is to put a small bright pen torch near the white part of the eye (while being careful not to poke yourself).

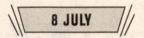

8 JULY

Why are eggs egg-shaped?

EGGS ARE EGG-SHAPED as a consequence of the egg-laying process in birds. The egg is passed along the oviduct by peristalsis – the muscles of the oviduct, which are arranged as a series of rings, alternately relax in front of the egg and contract behind it.

At the start of its passage down the oviduct, the egg is soft-shelled and spherical. The forces of contraction on the rear part of the egg, with the rings of muscle becoming progressively smaller, deform that end from a hemisphere into a cone shape, whereas the relaxing muscles maintain the near hemispherical shape of the front part. As the shell calcifies, the shape becomes fixed, in contrast to the soft-shelled eggs of reptiles which can resume their spherical shape after emerging.

There are many advantages to an ovoid egg. First, it enables the eggs to fit more snugly together in the nest, with smaller air spaces between them. This reduces heat loss and allows best use of the nest space. Second, if the egg rolls, it will roll in a circular path around the pointed end. This means that on a flat (or flattish surface) there should be no danger of the egg rolling off, or out of the nest. Most eggs will roll in a curved path, coming to rest with the sharper end pointing uphill. There is in fact a noticeable tendency for the eggs of cliff-nesting birds to deviate more from the spherical, and thus roll in a tighter arc. Third, an egg shape is more comfortable for the bird while it is laying (assuming that the rounded end emerges first), rather than a sphere or a cylinder.

Finally, the most important reason is that hens' eggs are the ideal shape for fitting into egg cups and the egg holders on the fridge door. No other shape would do.

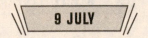

9 JULY

Can I become a fossil?

SO YOU WANT to become a fossil? This is admirable, but you have made a bad start. A hard, mineralised exoskeleton and a marine lifestyle would have given you a better chance. But let's start with what you have got: an internal skeleton and some soft outer bits.

You can usually forget the soft bits. If you really want to survive the ravages of geological time then you need to concentrate on teeth and bones. Fossilisation of these involves additional mineralisation, so you might want to get a head start and think about your diet: cheese and milk would build up your bone calcium.

After that it comes down to three things: location, location, location. You must find a place to die where you won't be disturbed for a long time, such as a cave. Alternatively, you need a rapid burial. Not a speedy funeral service, but something natural and dramatic – the sort of thing that is preceded by a distant volcanic rumble and an unfinished query along the lines of 'What was . . .?'

You might want to travel to find the right natural opportunity. Camping in a desert wadi in the flash-flood season would be good. And long walks across tropical river flood plains during heavy rain could get you where you want to be: buried in fine, anoxic mud. Or how about an imprudent picnic on the flanks of an active volcano? But take geological advice because you are looking for a nice ash-fall burial, not cremation by lava.

Talking of picnics, fossil stomach contents can provide useful palaeo-diet information, so a solid final meal would be good. Pizzas or hamburgers won't last, but shellfish or fruit with large seeds (you will need to swallow these) could intrigue future scientists.

Finally, trace fossils (marks in rocks that indicate animal behaviour) are always welcome. So a neat set of footprints leading to your final location would be good. Use a nice even gait with no hopping or skipping to confuse analysis of how you really moved.

Of course, you have more chance of winning the lottery than ending up as a fossil. But if you do go for a place in the fossil record please keep in touch. Geologists are always on the lookout for interesting new specimens, so let us know where you'll be. We can arrange to dig you up in, say, a million years.

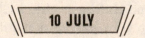

What would gravity feel like on a journey to the centre of the Earth?

THIS PROBLEM PIQUED the curiosity of no less a physicist than Isaac Newton, who solved it in his *Principia* (Book 1, theorem 33). If you are at the centre of Earth you are pulled equally in all directions, so you are in fact weightless. Higher up, at radius R from the centre, Newton found that the attractions of the materials in the hollow spherical shell of radius greater than R will all cancel one another out – a beautiful mathematical consequence of the fact that gravity decreases as the square of the distance between the objects increases. You feel only the pull of the mass in the sphere below you.

Newton showed that Earth's combined pull is simply proportional to the inverse square of the distance R from the centre. The mass of this sphere is proportional to its volume, that is, R^3. So, the weight you would feel, if you were foolhardy enough to descend through a homogeneous planet, would decrease in direct proportion to R^3/R^2 (which is equal to R) as you moved inwards, reaching zero at the centre.

In fact, the central parts of Earth – mostly dense iron – are much more massive than the outer parts, so your weight would decrease a bit more gradually at first and more rapidly as you penetrated the core.

Incidentally, if you could slide down a frictionless tube through the centre to your antipodal (opposite) point on Earth's surface and back, the round-trip would take you 90 minutes – exactly the same time as it takes to go round Earth in a low orbit.

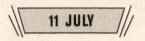
Why do teabags float?

WHEN YOU POUR boiling water over a teabag, it immediately rises to the surface. The first cause of the tea bag's inflation is likely to be the expansion of air inside the bag as it is heated from room temperature (around 25 °C, or 298 kelvin) to the boiling point of water (100 °C, or 373 kelvin).

According to Charles's law, the volume of a gas is proportional to its absolute (or kelvin) temperature. So the air in the tea bag will expand by a factor of 373 divided by 298, or roughly 1.25, a 25 per cent increase in volume.

A second reason for the inflation could be the phenomenon known as nucleate boiling. At atmospheric pressure water boils at 100 °C, but it is actually quite difficult for bubbles of water vapour, or steam, to form in the bulk of a liquid. Boiling usually occurs only at a solid surface where small cracks and crevices facilitate the formation of bubble nuclei, which then detach and grow as they rise. This is why in a pot of boiling water you will usually see bubbles appearing only at the base or walls.

What this means is that although the bulk of the water is super-heated and ready to boil, it is unable to do so until it comes into contact with a rough surface. The leaves in the tea bag would provide ideal nucleating sites for bubbles of steam to form, which would also help to inflate the tea bag.

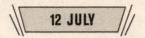

Why is the sea blue?

SEAWATER APPEARS BLUE because it is a very good absorber of all wavelengths of light, except for the shorter blue wavelengths, which are scattered effectively. The effect is caused by the selective absorption of light by water molecules, chiefly the oxygen component, that take out the red end of the visible light spectrum. In a similar way, ice masses at the poles and big icebergs look blue.

Reflection of light contributes to the colour of the open sea, but does not determine it. Even pure water is slightly bluegreen, because it filters out the red and orange content of light. However, impurities in seawater, especially organic substances, affect its appearance far more drastically.

Changes in the sea's colour are primarily due to changes in the type and concentration of plankton. Tropical oceans are clear because they are lacking in suspended sediment and plankton, which contrasts with the popular misconception that tropical waters have a high biological productivity. In fact, they are virtually sterile compared with the cooler, plankton-rich temperate ocean regions. Inorganic particulates and dissolved matter also reflect and absorb light, which affects the clarity of the water.

Blue Lake near Mount Gambier in South Australia is always blue, sun or no sun. The lake is situated in a limestone area and is saturated with calcium carbonate. The colour comes from the greater scattering of blue light by very fine particles of the compound suspended in the water.

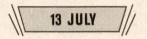

13 JULY

Why do some animals sleep lying down, while others sleep standing?

THERE ARE SEVERAL factors that affect an animal's sleeping style: how easily it can get up and stay up, how easily it can lie down, how comfortably it can stay down and what kinds of threats it faces in its environment.

In general, the larger a land animal, the less it likes getting up or staying down, so very heavy animals such as elephants don't lie down much. It is hard for recumbent elephants to breathe or to rise suddenly. Instead, their legs are adapted to support their weight vertically, so that a healthy elephant can stand indefinitely and sleep lightly, possibly leaning against something. It will only lie down for a deep sleep when it feels secure enough.

Giraffes sleep even less; they are more vulnerable to predators. A cat, pig or goat can take off quickly from a sleeping start, so they are free to lie down and curl up. Cattle and horses are somewhere near the break-even point. They keep to their feet as a rule, but can get tired enough to rest, or nap while lying down. In the wild, however, they rarely lie down when they are nervous of predators.

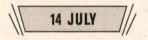

14 JULY

What language do deaf people think in?

IT VARIES. DEAF and hearing children raised by deaf parents who use sign language will acquire that language in a similar way to the acquisition of speech by hearing children. They will use this language to think with just as any other child will.

However, most deaf children are born to hearing parents who cannot sign, and they will not develop language in the same way. These children often use gestures for basic communication but this is not the same as sign language, and their first exposure to an accessible language may be at school, or later, and this is clearly disadvantageous.

Language is crucial to cognitive development, but what language is used and which modality it is in really doesn't matter. The idea that thoughts necessarily come in the form of words is also a misconception, especially common in the English-speaking world, perhaps because most people are monolingual there. Polyglots often report not thinking in words, but rather concepts. Thoughts only get put into words, or signs, when you want to communicate them or when thinking about communicating them.

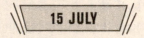

15 JULY

Can you measure the speed of light at home?

THIS IS AN amazing experiment that actually allows you to measure one of the fundamentals of science – the speed of light – in your own home. All you need is a long bar of chocolate, a ruler and a microwave oven.

Remove the turntable from your microwave – the bar of chocolate needs to be stationary. Put the chocolate in the microwave and cook at high power until it starts to melt in two or three spots – this usually takes about 40 seconds. You should stop after 60 seconds maximum, for safety.

Because the chocolate is not rotating, the microwaves the oven produces are not evenly distributed throughout the bar and spots of chocolate will begin to melt in the high-intensity areas or 'hotspots'. The microwaves race through the air at the speed of light in a wave-like fashion of peaks and troughs. The number of waves per second, also known as the frequency, is the key here. A standard oven will probably have a frequency of 2.45 gigahertz (the figure should be given on the back of the oven or in the instruction manual). If your oven is 2.45 GHz, this means the microwaves oscillate 2,450,000,000 times a second. So now you know how frequently the waves go up and down, you just need to know the wavelength. That will help you to calculate how fast they are travelling.

This is where the chocolate comes in. Because the wave will melt the chocolate when it goes through it on the upward part of the wave and the downward part, the distance between the globs of molten chocolate is half the wavelength of the microwaves in your oven. So double the measurement you have taken of the gap between the molten globs to find the microwave wavelength. In the *New Scientist* microwave oven, the distance between the globs of molten chocolate was 6 cm, so the wavelength in our 2.45 GHz oven is 12 cm.

To calculate the speed of light in centimetres per second you need to multiply this wavelength by the frequency of the microwaves. So you do 12 × 2,450,000,000 = 29,400,000,000, which is astoundingly near to the true speed of light of 29,979,245,800 cm per second (or 299,792,458 m per second as it is usually expressed).

If your chocolate bar is chilled beforehand, the molten areas tend to be more distinct when they first appear. Of course, you may find that a variety of different chocolate bars, all of which taste delicious slightly melted, will aid your research. True scientists know that it is always important to double-check results.

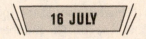

Could Jurassic Park happen in real life?

IN THE FILM *Jurassic Park* and its sequels, DNA from the stomachs of mosquitoes stuck in amber was used to recreate the dinosaurs. Could you do this in real life?

A mosquito found encased in rock in Montana is reckoned to have lived 46 million years ago, and another blood-feeding bug discovered in China predates that mosquito by 30 million years. This falls within the early Cretaceous, between 142 and 64 million years ago, which is when *Tyrannosaurus rex* and most other well-known dinosaurs lived, making their coexistence with blood-sucking insects highly likely.

The conclusion that these organisms fed on other animals' blood came from analyses of their abdominal contents. These identified iron-containing haem – the main component of blood, which carries oxygen and gives red blood cells their colour. The blood meal would have contained mainly red blood cells and some white cells. In mammals, red cells lack a nucleus containing DNA. All other vertebrates have red blood cells with a nucleus, each one containing the whole genome. So the DNA content would be much higher in a blood meal from a dinosaur than in one from a mammal.

However, the high level of degradation meant that no whole blood cells were identifiable in the recovered fossil material. Both digestion of the blood and fossilisation of the insect would almost certainly have destroyed any DNA. In addition, DNA has a half-life of approximately 500 years, and would be too broken to contain any information after about 1.5 million years. So finding a complete, intact genome in a dinosaur fossil or in an insect that fed on it is highly unlikely.

Let us imagine, however, that somehow an insect was embedded in amber just after sucking blood from a dinosaur, and preserved in super-ideal conditions. The owner of the extracted DNA could only be pinpointed by sequencing it and matching it to a reference genome. But we have no reference genome for a dinosaur, so there would be no option but to try it and see what comes out.

For that, we would need a whole, undamaged nucleus to transplant by injecting it into, say, an unfertilised frog egg with its own nucleus removed. The frog egg would provide everything necessary for development, and the transplanted nucleus would provide the genetic instructions for the new organism.

What happens then? A frog is an amphibian and does not make hard-shelled eggs, so would the transplanted dinosaur nucleus develop? Who knows? The Jurassic Park movies may have brought back to life the most amazing animals that ever walked the earth, but I wouldn't recommend trying it at home.

17 JULY

How do pebbles skim on water?

IN A PAPER published in the journal *Nature* in 2004, the French physicist Lydéric Bocquet and his collaborators revealed some of the secrets of successful stone skimming. They found that the optimum angle of attack is 20 degrees. So, even when the stone is thrown horizontally, the leading edge should be 20 degrees higher than the trailing edge. This maximises the number of jumps by limiting the contact time between the stone and the water, which is proportional to the energy dissipated.

The thrower also imparts spin to the pebble, providing a gyroscopic effect that stabilises its flight and preserves the original angle of attack when it bounces. In the absence of spin, the water would impart a torque on the stone and, because the trailing edge is the first to make contact with the water, this would tend to make it tumble.

The actual physics of stone skimming is not yet perfectly understood. However, the bounce could be understood as a result of the conservation of momentum and Newton's third law: when the stone exerts a force on the water, the water exerts an equal and opposite force on the stone. This lifting force is proportional to the density of the water, the surface area that is wetted and the square of the forward speed of the stone. Also, the bow wave created ahead of the stone when it strikes the liquid might act like a waterski jump, helping to launch the next hop. This minimises the contact time between the stone and the water, which in turn maximises the number of jumps.

Although ensuring the optimal angle of attack as the stone strikes the water, and imparting just enough spin to maintain stable flight are important, there are other factors. Selecting a flattish stone and having a fast throwing arm are examples.

Given that the urge to skim stones has been with us for thousands of years and the rules – getting the greatest distance or number of bounces – have remained unchanged since the ancient Greeks, perhaps this should become an Olympic sport. In the meantime, the current world record stands at 51 skips, set by Russell Byars in Pennsylvania on 19 July 2007.

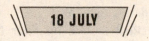

Can two galaxies collide?

THE UNIVERSE OVERALL is expanding. But gravity makes sure not all matter is travelling away from the centre. Look at the Earth as it orbits the sun. Half our time is spent travelling away from the centre of the universe (wherever or whatever that is), and the other half travelling back.

The big bang was not like a normal explosion, in which fragments of a lump of matter are blown out. Rather, the big bang set space itself expanding.

A common cosmological analogy is to think of galaxies as paper dots on the surface of a balloon. As the balloon inflates, the galaxies move apart, because the very space between them grows. In this analogy, it is the surface of the balloon, not the volume within, that represents the three-dimensional universe.

Galaxies may have their own trajectories across the surface of the balloon, pulled about by the gravity of other galaxies. This local movement is distinct from the expansion of space itself and means galaxies can collide. The Andromeda galaxy is actually moving towards us.

How much Neanderthal are you?

IN 2010, A group of geneticists announced that they had cracked the Neanderthal code. Using tiny fragments of 38,000-year-old DNA painstakingly pulled out of ancient bones, they had, against all odds, reconstructed a Neanderthal genome.

The geneticists decided to line up their Neanderthal genome against that of modern, living humans, and that revealed something amazing: there were bits that matched. Not just in the way that human DNA largely matches chimp DNA – when they looked closely, the team found that bits of modern human DNA were, in a sense, Neanderthal. The only explanation was that tens of thousands of years ago, a Neanderthal and a human had got down to it, had a hybrid child, and that child grafted fragments of Neanderthal DNA into the human family tree.

Today we know that 2 to 4 per cent of non-Africans' DNA is Neanderthal. The interbreeding happened after *Homo sapiens* left Africa, so people who have a very pure African lineage don't have any Neanderthal DNA. And we know that people didn't all inherit the same bits – in total about 30 to 40 per cent of the Neanderthal genome is still knocking about, spread out higgledy-piggledy in millions of living humans. If you have pale, freckly skin and red hair, there's a chance that the genes responsible were originally inherited from Neanderthals. You may also owe some of your immune system to them, and there's one gene people got from Neanderthals that governs the size of the tiny blind spot in our vision.

Of course, you could also flip the question on its head and ask how much of you is 'authentically human'. Which are the bits that we have and Neanderthals do not? There's a lot in there, of course, but one suite of genes is intriguing. They are interesting because they have previously been associated with domestication: dogs have them but not wolves, cats have them but not wild cats, and they are even found in a colony of Siberian foxes that have been domesticated, but not in their wild relatives. These genes give pets small skulls and narrower faces than their wild counterparts, and also make them less aggressive. So here's the lesson: many people are part

Neanderthal. But what really sets us apart from them, and probably all our ancestors, is that we are the domesticated species.

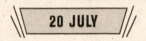

20 JULY

How do aquatic mammals see underwater?

HUMANS CANNOT SEE clearly under water without goggles. So how do aquatic mammals solve this problem?

For light reflecting off an object to be perceived as anything more than dim diffuse illumination, it must be focused on a single point on the light-sensitive retina at the back of the eye. The divergent light rays that strike the front of the eye must therefore be bent (refracted) to varying degrees in order to form an image.

Light is refracted when its waves cross at a glancing angle from one medium to another with a different refractive index. In terrestrial vertebrates light is refracted mainly by the curved surface of the cornea whose refractive index is considerably higher than that of air. The eye's lens has a similar refractive index to that of the surrounding parts of the eye and is responsible only for around one-third of the refractive power of the human eye, serving mainly to adjust the fine focus of the image seen.

Underwater the cornea becomes ineffective as its refractive index is very close to that of water. The underwater world becomes very blurry because light is focused a long way behind the retina and we become in effect very long-sighted. This can be rectified by putting air back in front of the cornea with a face mask or a pair of swimming goggles.

The same obviously cannot be true for animals that live underwater because otherwise their eyes would be of little use. Animals such as fish, cephalopods and aquatic mammals overcome the loss of a refractive cornea underwater by possessing more powerful spherical lenses that can deal with this problem, unlike the lens in the human eye. Next time you eat a fish take out the lens and you will see it is shaped like a marble.

21 JULY

Is glass a liquid?

TOUR GUIDES IN old European churches and cathedrals sometimes peddle the myth that medieval windowpanes are thicker at the bottom because of the slow flow of glass over centuries, suggesting that glass is a liquid. Unfortunately, glass is a solid, albeit an odd one. (And if you're wondering about the windows, it's because of the uneven way molten glass was originally rolled into sheets in the Middle Ages.)

Glass is called an amorphous solid because it lacks the ordered molecular structure of true solids, and yet its irregular structure is too rigid for it to qualify as a liquid. In fact, it would take a billion years for just a few of the atoms in a pane of glass to shift at all.

However, not everything about glass is quite so clear. How it achieves the switch from liquid to amorphous solid, for one thing, has remained stubbornly opaque. When most materials go through this transition between liquid and solid states, their molecules instantly rearrange. In a liquid the molecules are moving around freely, then snap! – they are more or less locked into a tightly knit pattern.

But the transition from the glassblower's red-hot liquid to the transparent solid we drink from and peer through doesn't work like that. Instead of a sudden change, the movement of molecules gradually slows as the temperature drops, retaining all the structural disorder of a liquid but acquiring the distinctive physical properties of a solid. In other words, in all forms of glass we see something unusual: the chaotic molecular arrangement of a liquid locked in place.

The process underlying this strange behaviour remains an open question. One possibility is that it's all down to energy use. According to the laws of thermodynamics, which govern how energy is transferred within a system, every collection of molecules is driven to find an arrangement with the lowest possible energy. But within any given system some patches do better than others, meaning different groups of molecules settle into different configurations – and, overall, into an irreconcilably chaotic arrangement.

But even if we put it down to thermodynamic laws, it's not clear what exactly drives the strange behaviour of glass. The push for low energy might be the prime mover. Then again, it could be the irrepressible tendency towards a maximum state of disorder. That's a perfectly plausible proposal, though it raises the troubling question of how ordered solids manage to survive.

Some researchers believe that glass may form in a manner not all that different from crystals, which have proved an easy target for analysis thanks to their repeating geometric structures. If they're right, maybe glass will finally become crystal clear.

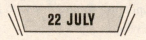

22 JULY

Why are pineapples spiky?

WHY DID PINEAPPLES evolve a fearsome array of spiny leaves that make the large, sweet and juicy fruit almost impregnable?

The short answer is that pineapples aren't eaten in the state that we normally see them but after they have ripened much further and fallen to the forest floor. The plant is a herbaceous perennial and grows up to 1.5 metres high and 1 metre wide. It has a rosette of long pointed leaves around a terminal bud. This bud produces the flowering stem which turns out an inflorescence of reddish purple flowers, each attached to the rest of the plant by a leaf-like structure called a pointed bract. In the wild these flowers may be pollinated by hummingbirds and will produce small, hard seeds in the fruit.

The pineapple fruit is created by the fusion of between 100 and 200 individual fruitlets that are embedded in a fleshy edible stem. The ovary of each flower becomes a berry, and all the berries coalesce into one solid structure. This is referred to as a multiple fruit or sorosis. The tough, waxy impregnable skin still contains the pointed bracts and the remains of the flower.

Although the pineapple plant can grow from seed, it also spreads very efficiently by a variety of vegetative means: from slips that arise from the stalk below the fruit, suckers that originate at the leaves, crowns that grow from the top of the fruits and ratoons that come out from the underground portions of the stems.

The pineapple we buy today in the supermarket is very different from its natural relatives in South America. The wild pineapple is much smaller. By the time it has dropped off its stem, hit the ground from quite a height and lain on the forest floor for a few days in the hot sun, it is very ripe and very soft. So when eaten, it is likely to be mushy and to split open easily, revealing the sweet and juicy fruit inside – very appealing to animals who will then disperse the pineapple's seeds elsewhere.

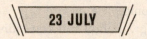

Why do tumble dried clothes feel softer?

To DO YOUR bit for the planet, you might be trying to use the tumble dryer less, or not own one at all. While hanging laundry outside dries clothes perfectly well, you might find that thicker materials, such as bath towels and socks, are hard and abrasive, in comparison to the soft and fluffy feel of the tumble-dried versions.

What is happening to the heavier fabrics in the tumble dryer is 'felting'. This is due in part to the centrifugal force applied by the action of the dryer, and also in part by the capillary attraction between fibres as the water drains away. When air-drying, if the water items are washed in is hard, then the minerals it contains will precipitate out, which helps to cement the fibres together.

To counteract this, you have to simulate the action of the tumble dryer in separating the fibres before drying takes place. The real experts in this are sea otters (though they don't have to deal with spin or tumble dryers). Watching videos of them fluffing up their own fur and that of their cubs might be instructive – or at least entertaining, with a high 'aww' factor.

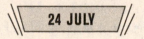

Why do fingernails grow faster than toenails?

ACCORDING TO LINDEN Edwards and Ralph Schott, who published a paper on the subject in 1937 in the *Ohio Journal of Science* (vol. 37, p. 91), toenails grow at half the rate of fingernails. On average, fingernails grow a little less than 4 centimetres a year – about the same rate that a tectonic plate moves!

There is quite a big variation between individuals, depending on heredity, gender, age and how much they exercise. Nails also grow faster in the summer.

The explanation as to why fingernails grow slower than toenails is quite simple when you think about it. Toenails are subject to less wear and tear and so do not need to grow as quickly. Hands, however, are used much more often. Fingernails in particular can be useful to prise things open.

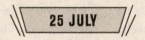

25 JULY

Why do bruises change colour?

A BRUISE OCCURS when small capillary blood vessels break under the skin. The haemoglobin in this leaked blood gives the bruise its classic red-purplish hue. The body then ropes in white blood cells to repair the damage at the site of the injury, which causes the red cells to break down. This produces the substances that are responsible for the colour changes.

The breakdown products of haemoglobin are biliverdin, which is green, and then bilirubin, which is yellow. Later, the debris at the bruise site clears and the colour fades.

It is the same process that disposes of red cells past their use-by date. White cells called macrophages break down defunct red cells in the spleen, liver, bone marrow and other tissues. Bilirubin is taken up by the liver, where it is converted to bile and used in the digestion of food.

It is bilirubin that helps to give faeces their characteristic colour. An accumulation of excess bilirubin in our body can occur in medical conditions such as hepatitis, giving the skin a yellow tinge also known as jaundice. One can sometimes observe this in some newborn babies. Jaundiced skin will itch because bilirubin is an irritant, while bruises are tender to touch. Ultraviolet light helps in breaking down bilirubin and is also the treatment for jaundiced babies.

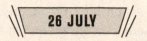

How did the Romans express fractions?

IN SCIENTIFIC WORK, which was always written in Greek, the Romans used fractions with a denominator of 60 (also known as sexagesimal fractions), like those used for angles and time, expressed by Greek literal numbers and positional notation.

For everyday uses, common fractions were always used, often spelled out, such as *tribus duas partes* ('with thirds two parts') for $\frac{2}{3}$. The solidus (the line in a fraction between the two numerals), or any implied division, was never used. However, abbreviations were often used if necessary. The most common were S or SK (*semisque*, or 'and a half') for $\frac{1}{2}$, and T or TK for $\frac{1}{3}$; $1\frac{1}{2}$ could be written ISK. A symbol for the sestertius (a unit of money worth $2\frac{1}{2}$ bronze asses), was IIS, later written HS. F, Z or FZ was used for $\frac{2}{3}$, while $\frac{1}{4}$ was represented by the usual modern division symbol ÷, or G, or a backward C. An overhead horizontal bar meant either $\frac{1}{12}$ or $\frac{1}{16}$, and could be combined with dots and other symbols. Common fractions were often expressed as sums of simpler fractions, for example $\frac{9}{16} = \frac{1}{2} + \frac{1}{16}$, written S- -.

A good place to see such fractions is *Book X* of Vitruvius's *De Architectura*, where they're used in connection with the construction of military machines. However, ignorant medieval (and modern) copyists have corrupted many fractions in existing texts because they did not understand the unfamiliar symbols. There are also a few examples of fractions in inscriptions but evidence is scanty and, as classical scholars usually pay little attention to numbers, science or mathematics, the usual reference works provide no enlightenment.

Roman practical calculations were done on the abacus, which used a decimal notation for whole numbers, but not for fractions (which were usually monetary fractions), and these had special columns of beads.

Incidentally, the notation for large numbers was different from that used in today's Roman numerals, and more convenient. Except for I, V and X, Roman numerals were not the usual alphabetical letters we see today, but symbols derived from Etruscan and other sources. L, C, D and M evolved from these, but many of the more unusual symbols remained for fractions. Even Greek letters are sometimes found. X was also used for $\frac{1}{16}$.

So, except for the simpler fractions, there was little standardisation. Roman engineers relied more on analogue and graphical methods of calculation than we do today, and numerical calculations were avoided as far as possible.

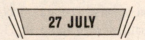

27 JULY

Why do dogs yowl at sirens?

WHEN EMERGENCY SIRENS pass by, you might hear a dog yowl – a cat, on the other hand, is likely to ignore it.

The reason dogs yowl when emergency services go by may be because, to the dogs, the siren sounds like other dogs howling and they respond by howling back. This goes back to the time when they hunted in packs and signalled to one another when searching for prey. Even if the siren does not mimic exactly the sound of another dog, they can probably pick out a component part of the screaming siren that does. Cats, on the other hand, hunt alone, are not pack animals and so do not respond to the sirens.

The same phenomenon occurs when families who attempt to sing music together are sometimes helped, or hindered, by their dog, which joins in when its human family breaks into a group howl.

Dogs, wolves and humans evolved as cooperative hunters, and more recently, sheep guardians, with a need to keep in touch with their partners on the next ridge. Hence howling, yodelling and such devices as the Israeli *challil*, or shepherd's flute. Sirens are artificial, amplified howling. Their rise and fall is calculated to alarm and stir you and to a dog's ears, they succeed splendidly.

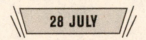

28 JULY

Why do we perceive the sun as yellow?

WHY DO WE perceive the sun as yellow? Because the sky is blue.

The sun's light is white, but when it hits the Earth's atmosphere it is scattered by the atoms, molecules and dust in the air. If this didn't happen the sky would be black, as it is in space. This was first explained by John Strutt, who became Lord Rayleigh, which is why the process is known as Rayleigh scattering.

Shorter wavelengths, such as violet and blue, are scattered more than longer ones. Because blue light is scattered across the entire sky – turning it blue – the remaining direct light from the sun appears richer in longer wavelengths such as yellow and red. When the sun is very low in the sky its light passes through more of the atmosphere, resulting in more of the spectrum being lost to scattering, making it appear more orange or red.

On a cloudy day we see everything in its true colours because there is no direct light and all wavelengths are transmitted equally.

Our eyes automatically adapt to the colour balance around them so we don't notice this, but photographic colour film can't do that. It is adjusted for sunshine, so cloudy photos will turn out blue.

Our eyes can even adapt to tungsten artificial light, but to capture this in a photo we need to use a film balanced for artificial light. Of course, as people switch to digital photography, it is much easier to correct for these problems.

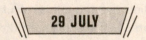

29 JULY

What is the fastest time an athlete can react to a starting gun?

A SPRINT ATHLETE is deemed to have false-started if they react within 0.1 seconds of the starting gun. But what studies have been done to test human reaction times, and is the fastest a person can react to the sound of a gun really exactly 0.1 seconds?

The earliest scientific research into human reaction times was undertaken in 1865, by the Dutch physiologist Franciscus Cornelis Donders, best known in his lifetime as an ophthalmologist but who was responsible for pioneering studies in what became known as mental chronometry.

Donders measured response times by applying electric shocks to the right and left feet of his subjects. They responded by pressing, as quickly as they could, an electric telegraph key to indicate which foot had been shocked. In some tests the subjects were warned beforehand which foot was to be tested, in others no prior notice was given. By measuring the difference in reaction times between the two types of test, which he found to be 0.066 seconds, Donders made the first tentative calculation of the speed of a human's mental responses.

Where a starting gun is used, the figure established by the International Association of Athletics Federations, 0.1 seconds, is in line with the response time measured by Donders, rounded up to one decimal place. So, yes, that is almost certainly the fastest time an athlete, even after repeated practice, could respond to the starter's gun. If they react any faster, the clear inference must be that they were launching themselves off the blocks before the gun was fired.

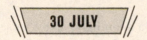

30 JULY

Why do most clouds have defined edges?

CLOUDS MAY APPEAR to be static objects. In fact they are dynamic and there is usually a convection current of air rising up into a cloud. The clouds with the most clearly defined edges are the billowing white cumulus clouds rising into a clear sky. These are formed by the condensation of water vapour as air expands and cools. This does not occur in a homogeneous layer but in a discrete parcel or column of warmer and less dense air rising from below through colder air above.

Although the cloud is cooled by expansion as it ascends, it continues to rise as long as its temperature is higher than that of the air surrounding it. Only when the air forming the cloud reaches a level at which the surrounding temperature is the same does it mix with that air and become fuzzy. Until then there is a sharp boundary between the different air masses.

At a given altitude, condensation occurs, thus defining a sharp lower boundary to the cloud. The sharpness of the upper surface depends on how fast the air is rising and the extent to which turbulence mixes this damp air with the surrounding, drier air. When the convection current ceases, the cloud will tend to become diffuse and lose its well-defined edge.

Glider pilots use the appearance of the clouds above them in order to recognise where there are thermal currents that will enable their planes to gain altitude.

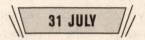

31 JULY

How are neon lights made?

NEXT TIME YOU find a neon light, take a closer look: you're looking at the stuff of stars. Most bright stars make neon and eject some of it from their coronae.

On Earth, practically all our neon comes from traces once trapped inside the solid mass of the planet and now escaping continuously through volcanism and sea-floor spreading. Neon atoms are too massive to diffuse rapidly out into space as helium does, so they loiter permanently in Earth's atmosphere, as do argon, krypton and xenon.

Neon amounts to less than 20 parts per million of the atmosphere – equivalent to about a cubic metre in a 30-storey office block. But even at that concentration, it is the sixth most abundant gas present in the atmosphere, and our most abundant noble gas apart from argon. Fortunately, it is not very difficult to extract: the liquid-air industry produces tonnes of neon in the same way it does the other noble gases – by fractional distillation. Neon distils out first because it is the most volatile of the atmospheric gases apart from hydrogen and helium, which do not liquefy under the usual processing conditions.

Once extracted, neon is concentrated by the same means that we are all best made to concentrate: by being put under pressure.

AUGUST

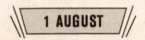

Why does newspaper tear nicely in one direction, but not another?

IT'S TRUE, NEWSPAPER looks homogeneous, but when you try to tear it from top to bottom or from side to side, there is a noticeable difference. And newspaper isn't the only everyday object that has a grain . . .

Most paper is made on machines which operate at high speed. The sheets are made by draining a dilute suspension of fibres and mineral fillers on a table of continuous and fast-moving, synthetic, sieve-like wire. The sheets are then consolidated by pressing and finally dried on heated cylinders before being put onto reels.

As the paper-making suspension is discharged from a vessel known as the flow box or head box onto the rapidly moving wire, most fibres – which are cylindrical in shape – are aligned in the direction of the wire's movement. This is called the 'machine direction'. The orientation of fibre in the paper structure allows the sheet to tear more easily in the machine direction than in the 'cross-direction' where the fibres are presented sideways.

The strength of all paper made in this way is influenced to varying degrees by the directional properties of the sheet. The addition of fillers and mechanical or surface treatment during manufacture tends to modify or reduce this. Newsprint tears more easily in the machine direction than computer or photocopier paper, as it contains mostly fibre and is lighter in weight.

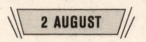

Why do animals have organs?

SURELY IT WOULD be more efficient and robust for us to have a thousand tiny hearts distributed round our bodies rather than one big one in the middle?

Insects and other small creatures have no hearts; instead they have a dorsal vessel that contracts in several segments to circulate haemolymph around the body. This fluid does not carry oxygen, however, because the gas exchange required for respiration can be accomplished by simple diffusion from the surrounding air. Bigger creatures such as ourselves need a more densely packed surface for gas exchange (such as gills found in fish or lungs in other animals), a network of blood vessels to transport the gas, and a pump to push the blood around.

The heart is evolution's best attempt at a high-pressure pump, and in terms of power-to-weight ratio it is likely to outperform a distributed system. However, this arrangement does introduce friction between the blood and the vessel walls. Therefore the blood pressure needed to operate a closed circulatory system increases with body mass. Taller animals must also force blood higher against the force of gravity.

The idea of having redundant hearts is nice, but for this to work each little pump would need to have its valves open if it fails, so that blood could still be pushed through it. To maximise efficiency each heart would have to beat when the pressure at its inlet was at a maximum, a far more complex task to manage than a single heartbeat.

What would you see if you took a compass into space?

THE MAGNETIC FIELD of the Earth looks like a dipole (the shape formed by iron filings around a bar magnet), although the Earth's is rotated about the field's axis to form a three-dimensional shape. This extends to about 60,000 kilometres into space. On the ground we use a compass in two dimensions. In space you can use a 3D 'compass' to map out the Earth's magnetic field, again giving an indication of north.

Beyond 60,000 kilometres into space, in the direction of the sun, we exit the Earth's magnetosphere and pass into the solar wind, which also carries the sun's magnetic field. During undisturbed solar periods the sun's field is shaped like a spiral, thanks to the sun's rotation, in the same way that a hosepipe whirled over your head emits a spiral of water.

Magnetic field measurements are made by interplanetary space-craft to understand how the sun's magnetic field and solar wind interact with the Earth's magnetic field. For instance, auroral displays are generated by the solar and terrestrial magnetic fields interlinking to allow solar wind plasma to enter the atmosphere.

On the opposite side of the Earth from the sun, the Earth's magnetic field is pulled into a long magnetic tail by its interaction with the solar wind, typically to 7 million kilometres or more. A compass in this geomagnetic tail would point along the tail, either towards or away from the Earth.

It is interesting to note that if we left the solar system beyond the 'heliopause' where the solar wind ceases to have an effect and travelled into interstellar space (approximately 150 astronomical units from Earth), our compass would start to measure the galactic field. Here, our magnetic field measurements might point towards the constellation Pyxis, appropriately better known as the Compass.

Does the ten-second rule work?

YOU'VE DROPPED A piece of food on the floor, but picked it up within seconds. Is it safe to eat? Childhood dictum says so, but what does the science say?

Sadly, this is an urban myth. Jillian Clarke is the youngest recipient of the Ig Nobel prize, won in 2004 for her study while still in high school of the 5-second rule. The time chosen for the 'rule' varies, but she traced its origins to at least as far back as Genghis Khan, when it was the 12-hour rule.

Clarke discovered that the quicker food is scooped off the floor, the fewer bacteria are transferred. Even so, while you would have to be unlucky to get ill, 5 seconds is long enough for food to be contaminated with a lethal dose of *E. coli*.

The number of bacteria that reaches the food depends on various factors: the population density of bacteria on the floor, the contact area between food and floor, and the presence of moisture. Not surprisingly, wet food collects more bacteria than dry food.

That's because, at the microscopic level, food leaves a tread mark because neither food nor floor is perfectly flat. This means that the two surfaces cannot mate perfectly, leaving gaps that bacteria cannot cross. However, if either or both the food or the floor is wet, moisture fills these gaps, allowing bacteria to swim to the food, effectively increasing the contact area. When either the food or floor is wet, there is also a risk that dirt on the floor will adhere to the food.

Whether or not you still eat it, is another matter. For any given food and surface combination, desirability and perceived contamination varies between people – everyone has a different 'yuck' point. Thus, for one person, 10 seconds on a living-room carpet may result in an errant chocolate button being binned, while for another, discovering one several days after it rolled under a table may result in a cry of delight followed by a quick brush off before it is popped in the mouth.

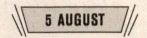

Why does an open car window on a motorway make so much noise?

IF YOU'VE ROLLED down the windows while driving along a motorway, you'll be familiar with the helicopter-like thumping noise that occurs, and gets louder the faster you go. What's going on?

An open sunroof or window essentially acts as an acoustics phenomenon known as the Helmholtz resonator. The most common example of this is the sound created when you blow across the top of an empty bottle.

The pressure inside the car resonates according to the cross-section of the opening, the speed of the air over it, and the volume of the cavity inside. These changes in pressure are perceived as a drumming sound. Imagine that the air inside and outside the car is separated by a thin invisible membrane: If something were to push down on the membrane and let go, the membrane would spring up and down with the amplitude of the oscillations getting smaller as with an elastic band.

But what if a downward push arrived just as the membrane was already moving downwards? This is where vortices come into play. They create resonance, maximising the amplitude by pushing down and increasing the pressure in the car just as the imaginary membrane is moving downwards anyway. This is similar to the way that people time their push on a child's swing. You push hard just as the swing is starting to move downwards again to get maximum thrust. The resonant frequency is shifted higher for bigger openings and smaller volumes, and vice versa. So blowing across a half-empty bottle should generate a higher frequency than blowing across an empty one.

6 AUGUST

What would happen if the Earth were hollow?

IF EARTH WERE hollow we would be in danger of death by suffocation, thirst, frying, starving, freezing and drowning, in that order.

A hollow Earth would not have enough mass to hold on to an atmosphere by gravity, and all the surface water would boil away. If the crust had enough mass to make up for the hollow centre, there would be no magnetic field, which is generated by Earth's liquid iron interior. Compasses wouldn't work, and some migratory animal species might get lost, but those would be the least of our worries because deadly radiation from the sun and outer space could then penetrate to Earth's surface.

If this could be solved, then presuming we could grow gills we could live underwater. We'd need to, because within a million years the continents would have eroded to little more than sandbanks, and the sea level would rise because of all the sediment dumped in the oceans by the rivers. It is only subduction of tectonic plates – where one plate moves under another – and mountain building, created by the same convection currents in the interior that create Earth's magnetic field, that keep uplifting the land to compensate for erosion.

Volcanic eruptions and subduction also play an important role in regulating the carbon dioxide levels in the atmosphere. On a hollow Earth without these processes, plant growth could cease entirely because of all the carbon reaching the ocean floor through erosion, and Earth would enter a period of deep freeze, deprived of the essential warming effect – and food supply – that carbon dioxide gives us now.

On the plus side, one could save time and money on flights from the UK to Australia by cutting holes on opposite sides of the globe and hopping through one of them – wearing protective clothing of course. The journey time would be a mere 7.5 hours.

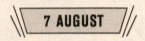

7 AUGUST

Do emotions evoke characteristic facial expressions?

Is THERE ONE expression everyone uses to convey shock, another for disgust, and so on? In 1924, Carney Landis, a graduate student in psychology at the University of Minnesota, designed an experiment to find out.

Landis brought subjects into his lab and drew lines on their faces with a burnt cork so that he could more easily see the movement of their muscles. He then exposed them to a variety of stimuli designed to provoke a strong emotional reaction. For instance, he made them smell ammonia, listen to jazz, look at pornographic pictures and put their hand into a bucket of frogs. As they reacted to each stimulus, he snapped pictures of their faces.

The climax of the experiment arrived when Landis carried in a live white rat on a tray and asked them to decapitate it. Most people initially resisted his request. They questioned whether he was serious. Landis assured them he was. The subjects would then hesitantly pick the knife up and put it back down. Many of the men swore. Some of the women started to cry. Nevertheless, Landis urged them on. In the pictures Landis took, we see them hovering over the rat with their painted faces, knife in hand. They look like members of some strange cult preparing to offer a sacrifice to the Great God of the Experiment.

Two-thirds of the subjects eventually did as they'd been told. Landis noted that most of them performed the task clumsily: 'The effort and attempt to hurry usually resulted in a rather awkward and prolonged job of decapitation.' Even when the subject refused, the rat did not get a reprieve. Landis simply picked up the knife and decapitated the rodent himself.

With hindsight, Landis's experiment presented a stunning display of the willingness of people to obey orders, no matter how unpalatable. That was something investigated in notorious obedience experiments conducted by Stanley Milgram at Yale University almost 40 years later, although his findings have since been questioned. Landis, however, never realised that the compliance of his subjects was potentially more interesting than their facial expressions. He remained single-mindedly focused on his research topic. And no, he was never able to find a single, characteristic facial expression that people adopt while decapitating a rat.

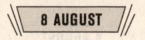

Why do rubber bands heat up when you stretch them?

THE NEXT TIME someone flicks a rubber band at you, don't shoot it back at them – at least, not straight away. First, take a moment to try a simple experiment which reveals a strange behaviour of this everyday object.

Stretch the rubber band between your hands and extend it as far as it will go. Now hold it against your upper lip and let it contract quickly. Wait a moment, then, keeping it against your lip, quickly extend it again. You'll notice that as you let the rubber band contract it feels cool. As you extend it the rubber feels hot. Quickly release the band and it will feel cool once more.

Rubber is made up of a tangled network of extremely long, flexible polymer chains – a bit like a crumpled fish net – that are constantly vibrating and rotating. Polymers are substances with molecular structures built up mainly from a large number of similar units bonded together.

As you stretch a rubber band, these polymer chains align and are pulled closer together, reducing the range of motions they can undergo. This reduces their entropy (the degree of disorder or randomness they possess). So, according to the laws of thermodynamics, to maintain equilibrium, entropy has to increase in some other way to counteract this, and the atoms in the chains compensate by vibrating more vigorously. As a result, the rubber heats up.

When the rubber is allowed to contract, the process reverses – the chains become free to rotate and vibrate, and their entropy or randomness increases. To compensate, the entropy of the atoms in the chain decreases, the atoms vibrate less and their temperature falls.

Now you can shoot the rubber band back . . .

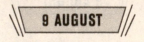

9 AUGUST

How do they make toothpaste stripy?

THIS IS ONE of those simple inventions that has been helping manufacturers to sell toothpaste for decades. We have to go back almost 50 years to find US patent number 2,789,731 and UK patent 813,514, both in the name of Leonard Lawrence Marraffino. He licensed his invention of striped toothpaste to Unilever, and this company subsequently marketed the first commercial version, which was called Signal in the UK. When the tube was squeezed it produced red stripes on white toothpaste.

Behind the nozzle was a hollow pipe that extended a little way back into the toothpaste tube. The white paste travelled down this pipe. Around this pipe was the funnel-shaped neck of the tube and, at the nozzle end of this, the pipe had tiny holes that opened into its interior. When filling the tube, red paste was first squirted into the funnel-shaped neck until just shy of the rear entrance of the pipe. Then the white toothpaste was added, and the end of the tube sealed.

When you squeezed the tube, the white paste flowed out through the pipe, but it also compressed the red paste, forcing it through the tiny holes into the pipe to form the stripes. If the stripes stopped coming out, they could sometimes be made to start again by warming the toothpaste tube in hot water.

Colgate-Palmolive's US patent number 4,969,767, granted in 1990, describes a scheme for adding stripes of two colours. Here, the central pipe is surrounded by a wider but shorter pipe, which creates a space for a second coloured paste. Tiny holes once again deliver this paste onto the surface of the white paste.

10 AUGUST

What would happen if you stood in the Large Hadron Collider?

AT LEAST ONE unfortunate person – Anatoli Bugorski – has already been struck by a high energy particle beam. In 1978 his head was briefly caught in the proton beam of a particle accelerator, the U-70 synchrotron in Russia. He reported seeing a flash of light 'brighter than a 1000 suns' but experienced no immediate pain. The full extent of his injuries only became apparent over the next few days – he lost half of the skin on his face where the beam had burned a path through it, and experienced various other complications.

To the surprise of doctors he survived, albeit with lifelong symptoms. It isn't clear whether the beam was operating at its full capacity of 70 gigaelectronvolts at the time of the accident. Presumably if one did the same thing at the Large Hadron Collider, which has a beam energy of 6.5 teraelectronvolts – almost 1,000 times greater and double again at the point where the beams collide – the aftermath would not be pleasant.

There is also the experience of Ray Cox, one of the North American victims of the infamous Therac-25 radiation therapy machine in the mid-1980s. As a result of poor hardware and software design, he was targeted with an electron beam that was more than a hundred times the intended dose. In his own words this felt like 'an intense electric shock' and he fled the treatment room screaming in pain. Some victims of the defective equipment died of radiation poisoning. The beam energy of this machine was a mere 25 megaelectronvolts.

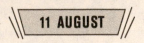

11 AUGUST

Why haven't we found a cure to cancer yet?

Cancer is a very complex disease, which makes it hard to develop treatments.

One complication is that cancer cells can grow and spread without symptoms for a long time. Hardly anyone dies from the initial, cancerous lump, called the primary. What kills people are the secondaries, fragments of the original lump that break off, spread to other parts of the body and start growing as new lumps, a process called metastasis. Often the secondaries spread without a person even knowing that they have a primary. By the time they feel symptoms, the secondaries are too numerous to treat through surgery.

When trying to develop treatments, the other major complication is that cancers originate from random cells in random organs in random people, so they're all completely different. And cancer cells are a mess, both genetically and in the way they behave. They cut loose from their normal job and go haywire, switching on genes that should be off and activating forbidden cellular processes, enabling them to multiply uncontrollably. It's less an abnormal cell or tissue you're trying to kill, more a constantly evolving ecosystem.

But thankfully, medicine is making gradual progress. The best solution once a cancer has been diagnosed is to cut it out surgically. Usually, patients also receive chemotherapeutic drugs that – unlike the surgeon's knife – can circulate throughout the body and pick off any secondaries before they become a problem.

Recently, researchers have been turning to a solution that aims to prime the immune system to fight the cancer. The beauty of this is that the immune system can evolve too, and so potentially keep pace with a cancer doing the same thing. These new drugs like nivolumab and pembrolizumab work by tearing up the deceptive cloak of the tumour and allowing the immune system to do the rest. These drugs have had profound impacts on the survival of patients with otherwise fatal skin cancers and lung cancer – the biggest cancer killer worldwide – helping some live for years rather than just a few months. So while there may never be a single 'cure for cancer', the future is looking bright.

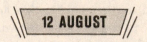

Why doesn't air separate out?

OXYGEN HAS A slightly greater density than nitrogen. So why don't these main constituents of air separate out?

A change in the ratio of oxygen to nitrogen would be expected in a hypothetical quiescent, or still, atmosphere. However, constant mixing occurs in the real atmosphere, driven by Earth's rotation and also by differences in density between hot air at Earth's surface and colder air higher up.

Up to altitudes of between 80 and 120 kilometres, this mixing results in a fairly uniform concentration of oxygen and nitrogen – which respectively make up approximately 21 per cent and 78 per cent of the atmosphere. This region is known as the homosphere. Partial stratification of the two gases does occur above 120 kilometres, in the heterosphere, where the density of air is much lower than at the surface and the efficiency of bulk mixing processes is reduced.

At room temperature, gas molecules move rapidly, with oxygen and nitrogen travelling at around 500 metres per second, so they obviously collide frequently. This allows the oxygen and nitrogen molecules to mingle and mix, rather like large numbers of people on a nightclub dance floor, in a process known as diffusion. Convection, the transfer of heat within the atmosphere, also plays an important role in gas mixing.

Gas mixing is a spontaneous process. This means that if you had a container with two compartments separated by a barrier, with one compartment containing pure nitrogen and the other pure oxygen, the two gases would automatically mix as soon as the barrier was removed. It is like entering a classroom full of teenage boys and girls on the last day before the summer break and patiently expecting them to settle down to serious work. It won't happen.

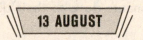

Why does a parachute have a hole in the top?

IF YOU FIND the thought of jumping out of a plane nauseating, the fact that there is a big hole in the top of the parachute probably doesn't make the idea any more appealing. But that hole – known as an apex vent – is crucial to a safe landing.

In the days before the apex vent, the only way that the air trapped underneath the parachute could escape was to spill out from one edge of the canopy, thereby tilting it and throwing the hapless parachutist to one side.

As the canopy swung back, more air would spill out from the opposite side, setting up a regular, pendulum-like oscillation (watch any footage of Second World War parachutists and you will see this).

As you can imagine, hitting the ground during a downswing was understandably hazardous, especially if it was also a windy day. The apex vent, by allowing the air to leak slowly out of the top of the parachute canopy, prevents this wild oscillation and makes for much safer landings.

Another benefit of the apex vent is that it slows down the opening of the parachute. Without the vent, air inflates the canopy much more abruptly, and it can damage the parachute or bring tears to the eyes of (particularly male) jumpers.

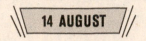
What do astronauts do with their dirty underwear?

ONE OF SPACE travel's most pressing but least-known problems is what to do with dirty underwear. One solution, attempted by Russian scientists in 1998, was to design a cocktail of bacteria to digest astronauts' cotton and paper underpants. They had hoped that the resulting methane gas could be used to power spacecraft.

The disposal unit would be able to process plastic, cellulose and other organic waste aboard a spacecraft. Cosmonauts identified waste as one of the most acute problems they face. Each astronaut produces an average of 2.5 kilograms of uncompressed waste a day. To keep waste to a minimum, they are forced to wear underwear for up to a week, and discarded undergarments are burned up in the Earth's atmosphere in waste modules that call twice a year. As for their poo . . .

From the outside, the 1972 space shuttle toilet looked much like the uncomfortable apparatus found in the jet airliners of the era. Inside, however, liquid and solid wastes were directed into separate pipes by high-velocity air streams that compensated for the lack of gravity on board the shuttle. The waste was held in two tanks – solids being vacuum-dried, sterilised and deodorised, ready to be pumped out when the shuttle returned to Earth. Depositing them in orbit would have obscured rearward vision and possibly interfered with external equipment.

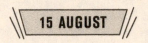
Why do fizzy drinks taste worse when they go flat?

A GOOD TASTE is a matter of blended, often contrasting, sensations and expectations. These include temperature – for hot and cold drinks, say – sound and texture for crisps or creams, plus aroma, flavour and stimuli on the tongue. A good fizz tickles the nose and splashes minute stimulating droplets around the mouth as you drink.

Most fizzy drinks are made so by injecting carbon dioxide into the liquid at high pressure. Carbon dioxide dissolves readily at atmospheric pressure, but the high pressure allows even more to be dissolved. It forms carbonic acid in the drink, and it is this which gives the drinks their appealing 'fizzy' taste – not the bubbles, as many people believe. When the drink goes flat, most of the dissolved carbon dioxide has been released back into the atmosphere, so the amount of carbonic acid is also reduced.

The fizzy taste is more appealing than the flat one simply because the drink was meant to be fizzy. Cola and champagne, for example, are concocted with the fizz in mind, using the carbonic acid as an essential ingredient in the flavour, so they will naturally taste better when the drink is still fizzy. Dissolved carbon dioxide has a distinct taste of its own, which is slightly sharp. Flat beverages have lost this bite. Going flat upsets the balance of the flavours and other stimuli, and without them such a drink is likely to taste insipid or too sweet, and . . . well . . . flat.

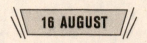
Why does your voice go hoarse when you shout?

PEOPLE WITH TRAINED voices, such as singers and actors, can often produce amazing volumes of sound without much effort and with very little harm or discomfort to their throats.

Roughly speaking they form their mouths and throat from the larynx upwards into an exponential horn, the most efficient shape for sound production. Suitably coupled with the vibration of air, a small amount of vocal energy can produce a formidable volume.

The untrained voice produces louder sounds by brute force rather than technique. We force air more violently between our vocal cords, thereby damaging them. Cords react largely by coating themselves with more and gummier protective mucus, and by fluid swelling. Both effects interfere with their correct, efficient vibration and stop them closing and opening in the proper way, causing hoarseness.

If you are lucky enough to escape bacterial infection in the damaged tissues, then a little rest – perhaps a day or two – will give the tissues time to recover. Should you persist in such abuse, you risk permanent injury, though your vocal cords will very likely become calloused and you will stop becoming hoarse.

However, if you are in a profession that demands continual shouting, taking a course of voice training could help. Apart from escaping injury, you will be able to make yourself heard with considerably less effort.

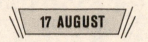
Does anything eat wasps?

THE LOWLY WASP certainly has its place in the food chain. Indeed, the question should possibly be 'what doesn't feed, in one way or another, on this potentially dangerous insect?'

Here are a few invertebrates that do: several species of dragonflies (Odonata); robber and hoverflies (Diptera); wasps (Hymenoptera), usually the larger species feeding on smaller species; beetles (Coleoptera); and moths (Lepidoptera).

The following are vertebrates that feed on wasps: skunks, bears, badgers, bats, weasels, wolverines, rats, mice and numerous species of birds. In fact, the definitive source on European birds, *Birds of the Western Palearctic*, lists a remarkable 133 species that at least occasionally consume wasps. The list includes some very unexpected species, such as willow warblers, pied flycatchers and Alpine swifts, but two groups of birds are well-known for being avid vespivores. Bee-eaters (Meropidae) routinely devour wasps, destinging them by wiping the insect vigorously against a twig or wire. And honey buzzards raid hives for food.

Last, but certainly not least, humans and probably some of our closest ancestors eat wasps too. You can eat the larvae of several wasp species fried in butter – we've heard they're quite tasty.

Could an Olympic high jumper escape the moon's gravity?

SADLY, EVEN AN Olympic high jumper would be unable to break free of the moon's gravity. However, they would be able to leap a lot higher than their earthly counterparts, because the moon's surface gravity is about six times weaker than Earth's. The current world record for the high jump is 2 metres 45 centimetres, therefore the moon record would be somewhere in the vicinity of 15 metres.

To escape the moon's gravity entirely you would have to jump at what is called the escape velocity. This is the speed you need to reach before you can entirely escape a body's gravitational field at a single kick. For Earth, this is about 11 kilometres per second. To escape the clutches of the moon's gravity you would need an escape velocity of 2.38 kilometres per second, about a fifth of the speed required to escape from Earth.

Escape velocity is the speed which is needed to escape with no further propulsion, but if you gave a slug a very long ladder and sufficient lettuce, it could keep slithering along all the way to infinity without ever having to travel at escape velocity. And so it was with the Apollo lunar modules, which didn't need to reach escape velocity to get back into orbit because they had an engine providing sustained power.

To fly over the surface of the moon would effectively require going into orbit (or simply leaping a very long distance). An orbit is really just a jump which never comes back down because the body over which you move is spherical and its surface is falling away from you as fast as you are falling towards it. In order to go into orbit just above the surface of the moon, a high jumper would need to leap at about 1 kilometre per second – still rather out of range.

So all the moon is really good for, if you are a high jumper, is setting new extraterrestrial records. You'll need to find a planet significantly smaller before you can fly off into the sunset. Those who can clock a respectable 11.4 seconds for the 100 metres on Earth would be running fast enough to launch themselves from the asteroid Toro: discovered in 1964, Toro has a radius of only about 5 kilometres. Those who are not quite as nimble on their feet might want to choose one of several smaller asteroids. With a radius of only 1 kilometre, a brisk walk on Geographus, for example, might do the trick. Its gravitational field strength is less than 1/6000th that of the moon.

19 AUGUST

What is the chemical formula for a human being?

ONE'S 'CHEMICAL FORMULA' depends on a number of factors, most notably whether we're talking about a he or a she. Male bodies contain more water than female bodies, which have extra lipids. By weight, oxygen amounts to about two-thirds of the body, followed by carbon at 20 per cent, hydrogen at 10 per cent and nitrogen at 3 per cent. Elements originating from pollutants would only be present in trace amounts.

If a human body were broken into single atoms, we would arrive at an empirical formula H_{15750} N_{310} O_{6500} C_{2250} Ca_{63} P_{48} K_{15} S_{15} Na_{10} Cl_6 Mg_3 Fe_1. The relative numbers of atoms in this differ from the composition by weight because atoms have different masses. However, metabolism, defined as the chemical and energy exchanges in a living body, means that any such chemical formula is continually changing.

Having a chemical formula for a process can be useful. If we find all the elements and determine all the mathematical expressions applying to them, the whole process can be determined. But this is not the whole story. Life is characterised by extensive, adaptive self-regulation of its own structural order, and utilises feedback control. An organism uses its resources in its own emergent way. The chemical reactions work, but how they are brought together is a matter of emergent control systems. This means that not only is it impossible to write an accurate formula for a human being, it is unnecessary and can be misleading to try. Life is what it does with chemical species, not just which ones it is made from.

20 AUGUST

If the stars are so bright, then why is the sky so dark?

STARS APPEAR AS dots of bright light in the night sky. As with our sun, every star represents an immense light source that is dazzling close up. So given the brightness of stars, why are the intervening spaces, when viewed from Earth, black?

The naked eye can see perhaps 9000 stars. Using medium-power telescopes this number grows to millions, and modern observatories may detect billions. Estimates vary as to the number of stars in the universe – or the observable universe, because what we see is limited by the speed of light. A minimum figure commonly cited is 10^{22}.

Why, then, are we not ablaze at night? This is what is known as Olbers' paradox, named after the 19th-century German astronomer Heinrich Olbers. At the time, the universe was thought to be static and set in an infinite sea of stars. Now we have considerably more knowledge about our cosmos.

The simplest, most widely accepted hypothesis is that many of these stars are so far away that their light has not yet reached us, and perhaps never will – at least not in a form we can see. The expanding universe delays or prevents light reaching us because of the distances involved, and maybe because galaxies recede from us at speeds effectively greater than that of light. And the wavelength of the light radiation lengthens as galaxies recede from our stand-point. This is known as red shift because the receding light source is shifted to the red end of the light spectrum.

Interestingly, the essence of this explanation is the same as that given by Edgar Allen Poe in his 1848 essay, *Eureka*:

> *Were the succession of stars endless, then the background of the sky would present us an uniform luminosity, like that displayed by the Galaxy – since there could be absolutely no point, in all that background, at which would not exist a star. The only mode, therefore, in which, under such a state of affairs, we could comprehend the voids that our telescopes find in innumerable directions, would be by supposing the distance of the invisible background so immense that no ray from it has yet been able to reach us at all.*

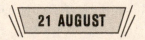

21 AUGUST

Can you improve your eyesight?

YOU CAN, OF course, improve your eyesight with glasses or surgery – but there's a neat trick to see things clearer with materials you can find at home.

Punch a tiny hole in a piece of card with a pin. Peer through the pinhole in the card with one eye while keeping the other eye closed. You'll notice that writing and objects far away become clearer through the pinhole. The effect is particularly noticeable if you are short-sighted (and aren't wearing your glasses). If you have excellent vision, the effect will not be as obvious.

When light enters the eye, it is bent by the lens at the front so it hits the retina at the back of the eye. This allows your brain to create a picture of what you can see. Under normal circumstances, light rays entering the lens are not focused in one place because light is coming from all directions. For you to see a clear image, your eyes have to concentrate all the rays into a single point on the retina.

If your eye is not perfectly shaped, as happens in people with short-sightedness, the outermost rays entering the eye are not bent by the lens enough for them to be focused on the correct point at the back of the eye. The innermost rays entering the eye do not need to be bent as much to hit the middle of the retina and these travel a relatively straight route to form a clear image, even in people who are short-sighted.

However, the outermost rays confuse and blur this image. By looking through a pinhole, the bundle of rays entering your eye is greatly reduced, because the hole allows through only the inner rays that pass through the central portion of your lens straight to your retina. It excludes the peripheral rays that cause the blurring. This means you see images clearly again (though they are darker).

Another thing you can try is using the gaps between the teeth of a comb to reproduce this effect. Native Alaskans have historically worn glasses with narrow slits in them, reproducing the effect of looking through a comb. More importantly, because snow and ice reflect a lot of light, looking through slitted glasses helps to reduce the amount of light entering the eye, aiding vision and preventing snow-blindness.

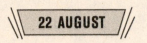

Is there a single foodstuff that a human could survive on?

ANY SINGLE SUBSTANCE such as water or fat? No.

Any single tissue such as potato? No.

But considering we must be allowed to continue drinking water and breathing air (even though those are also nutrients), we can relax our rules a bit. Not surprisingly, no strict monodiet can rival any healthily balanced diet, but there are two classes of foodstuffs that in appropriate quantities can maintain a reasonable level of health.

One such class is baby food. Some baby foods contain eggs, milk, certain seeds, and so on. Not as tasty as pizza and no single baby food is a perfect option, but some are adequate.

The other – and you might consider this a cheat – is whole animals. Oysters or fish such as whitebait or sardines might supply the necessary nutrient uptake. Animals sufficiently closely related to humans might also do, if eaten in the correct form and quantity. Farming families in the semidesert Karoo region of South Africa apparently ate mainly sheep or cattle.

For the most perfectly balanced human monodiet, however, other humans would be the logical food of choice. Not sure there would be many takers though – and we certainly aren't encouraging it!

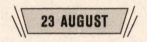

How do gnats survive heavy rainfall?

THE WORLD OF the gnat is not like our own.

A falling drop of rain creates a tiny pressure wave ahead (below the raindrop). This wave pushes the gnat sideways and the drop misses it. Fly swatters are made from mesh or have holes on their surface to reduce this pressure wave, otherwise flies would escape most swats.

Even if a raindrop were to hit a gnat, the gnat would probably still survive. Because of the difference in scale, we can regard a collision between a raindrop and a gnat as similar to that between a car moving at the same speed as the raindrop (speed does not scale) and a person having only one thousandth the usual density – for example, that of a thin rubber balloon of the same size and shape. A balloon is easily bounced out of the way, and would burst only if it was crushed up against a wall.

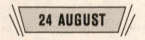

Does going on a plane inflate you?

IF YOU TAKE a packet of crisps on a flight, you'll notice that it will inflate. That's because cabin air pressure at 10,000 metres – the cruising altitude of modern airliners – is about two-thirds of normal sea-level pressure. The contents of any gas-filled elastic container will expand due to this difference in pressure.

What happens to that crisp packet also happens to your body.

Anybody who has flown with a heavy cold or with blocked sinuses may have experienced this effect, which can cause severe pain on ascending and descending as gases in the head expand.

Your head isn't the only place gases expand. Most of the gas in the gut is in the large bowel and if it expands, there is only one way for it to come out. Some find the liberation of colonic miasma one of the surprising pleasures of jet travel. Perhaps on learning this readers will never view fellow travellers with equanimity. Of course, this would almost certainly be hypocritical.

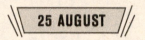

Why are swimming pools chlorinated?

SWIMMING POOLS ARE chlorinated to disinfect the water. But chlorine is not the only member of the halogen group that can be used for this purpose; iodine and bromine will also do the job, though not fluorine because it is too reactive. Chlorine is often chosen simply because it is cheap, readily available and relatively easy to handle.

Disinfection relies on disrupting a harmful organism's metabolism or structure. That can be achieved by oxidation and non-oxidising chemicals which have similar effects, as well as by non-chemical processes such as ultraviolet (including sunlight), X-rays, ultrasound, heat (as in pasteurisation), variations in pH and even storage to allow organisms to die naturally.

Chlorine gas consists of molecules of two chlorine atoms but no oxygen. When added to water, one of the atoms forms a chloride ion. The other reacts with water to form hypochlorous acid, an oxidising agent. Disinfection comes from the hypochlorous acid reacting with another molecule, most probably in the bacterial cell wall, in an oxidation-reduction reaction. If this happens enough times, the organism's repair mechanisms are overwhelmed and it dies. So concentration of disinfectant and the length of time pathogens are exposed to it are important factors.

Chlorine is available in many different chemical forms, such as chlorine gas, sodium hypochlorite powder (often used in home swimming pools), and chlorinated lime or bleaching powder. Some chemicals containing chlorine are not disinfectants because the chlorine in them, usually in the form of chloride, is completely reduced with no further oxidising power. Sodium chloride is such a chemical, which is why water cannot be disinfected using a pinch of salt, and why pathogens can survive in seawater.

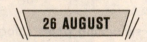

What causes freckles?

A FRECKLE CORRESPONDS to a higher concentration of the pigment melanin, and is most obvious when it contrasts with fair-coloured skin. Freckles are associated with variants of the gene on chromosome 16 for the melanocortin-1 receptor (MC1R), which are also responsible for red or ginger hair. This probably explains why there is a correlation between freckles and red hair.

Melanocytes in the skin produce melanin and package it into organelles called melanosomes. These are passed into overlying keratinocytes, the cells that form the outer barrier of our skin, where they release their payload of melanin. Those born with darker skin have larger melanocytes, which lead to more melanin in the outer skin cells. Freckles are also associated with bigger melanocytes.

Freckles are triggered by exposure to sunlight. UVB radiation activates melanocytes to increase melanin production, which can cause freckles to darken, increasing effectiveness as a sunscreen. The person tans relatively quickly where they have freckles, but the skin between is still prone to burning.

Red hair and freckles occur most frequently in people with northern or western European ancestry. For example, 13 per cent of Scots are redheads and about 40 per cent of them carry the red-hair gene.

Fair skin and freckles might bestow an evolutionary advantage to those living at high latitudes, where it is colder and the intensity of sunlight is lower. It is suggested that a paler complexion reduces heat loss through radiation, though clothing would surely be more effective at retaining heat. The lighter skin pigmentation between freckles also leads to greater absorption of sunlight and higher production of vitamin D, reducing the incidence of rickets in northern latitudes.

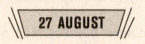

27 AUGUST

What happens when lightning strikes water?

WHEN LIGHTNING STRIKES, the best place to be is inside a conductor, such as a metal-hulled boat, or under the sea (assuming you are a fish).

When a bolt of electricity, such as a lightning bolt, hits a watery surface, the electricity can run to earth in a myriad of directions. Because of this, electricity is conducted away over a hemispheroid shape which rapidly diffuses any frying power possessed by the original bolt. Obviously, if a fish was directly hit by lightning, or close to the impact spot, it could be killed or injured.

However, a bolt has a temperature of several thousand degrees and could easily vaporise the water surrounding the impact point. This would create a subsurface shock wave that could rearrange the

anatomy of a fish or deafen human divers over a far wider range –
tens of metres.

If someone in a metal-hulled boat was close enough to feel the
first effect they would be severely buffeted by the second. Besides
which, metal hulls conduct electricity far better than water, so a
lightning bolt would travel through the ship in preference to the
water.

Last century, the physicist Michael Faraday showed that there is
no electric field within a conductor. He demonstrated this by climbing
into a mesh cage and then striking artificial lightning all over it.
Everybody except Faraday was surprised when he climbed out of
the cage unhurt.

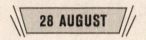

28 AUGUST

Why are soap bubbles multicoloured?

BLOWING BUBBLES IS fun for kids and adults alike – blow a bubble
with some soap and a bubble wand, and you can produce bubbles
of all sizes and colours. Look closely, and you may even spot that
the outer skin of a bubble is one colour, while the inside appears to
be another. If you're using the same liquid and wand, how is this
possible?

The physics of soap bubbles is a fascinating subject. The colours
are most commonly caused by thin-film interference between light
rays reflected from the outer and inner surfaces of the bubble.
Depending on the thickness of the bubble wall, certain wavelengths
of light will interfere constructively, giving rise to strong colours. As
the bubble evaporates, the wall thickness will change and hence so
will the colours.

Because the soapy water tends to flow downwards under gravity, the thickness of the bubble wall may also vary from top to bottom, giving rise to horizontal bands of colour. This can be reduced by thickening the soap solution, for example, by adding glycerol. When the bubble wall becomes very thin – thinner than the wavelength of visible light – the colour will disappear. These regions are transparent, but may appear dark because they are usually surrounded by coloured ones. The bubble will burst a few seconds after these patches appear.

Colours can also be produced by adding dyes to the soap solution. Most water-soluble dyes will not work, however, because of the thinness of the bubble walls, and because the water collects at the bottom of the bubble. The dyes need to bind to the soap used to make the solution. If you want the most spectacular bubbles, find somewhere like an ice rink – the cold air and high humidity will allow the bubbles to persist for up to a minute and they can show some quite spectacular colours.

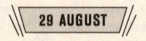

29 AUGUST

Why do phantom traffic jams occur?

IF YOU'VE DRIVEN down a busy motorway before, you've probably encountered an inexplicable standstill. The traffic will come to a halt, despite there being no accident or junction, and after some time start moving again. This is known as a 'phantom' traffic jam.

In open, free-moving traffic, each car is basically autonomous and can travel as fast as its driver wants to go. In denser traffic, there is interaction between vehicles, and if one car slows down then the one behind must also slow down.

When the traffic reaches a certain critical density, a 'shock wave' can spontaneously travel back through it. This is because when one driver brakes gently, the driver behind will choose to slow down more markedly. The effect becomes more pronounced as it works its way backwards. Meanwhile new traffic keeps on arriving at the same rate. With nowhere for it to go, it comes to a grinding halt. The front end of the jam gradually clears, and when cars at the back finally get moving again, the road in front of them is virtually empty, and drivers wonder what the problem was.

Traffic congestion is hard to model because it is very non-linear and dependent on human reaction times. If drivers had a reaction time of zero and could respond instantly to changes in the flow, an entire highway could start and stop together, like soldiers on parade. But we know from watching cars take off at traffic lights that there is roughly a 1-second delay between each car starting to move. It is probably no coincidence that the critical traffic flow that results in a complete standstill is about one vehicle per second.

To kill these waves in dense highway traffic, leave six to eight car lengths to the next vehicle, and as soon as you can see the brake lights of cars further along go on, take your foot off the accelerator and coast. If the car ahead of you regains speed, you can easily catch up, so avoiding a general slowdown. The cars behind will have slowed down only slightly while you coasted, so they can also speed up when you do.

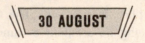

30 AUGUST

Is it possible to tell how old someone is without records?

AGE EVALUATION IS a growing area of involvement for forensic practitioners: it is often needed for individuals seeking asylum or refugee status, who may be genuinely unaware of their age. A requirement is placed on the practitioner to assign age with the greatest degree of accuracy possible, because there can be big benefits in being assigned a certain age.

The relationship between actual age and 'biological age' is strongest in the young. We can assign the age of a child to within a few years, but even here it is not an exact science, because environmental and genetic factors will have an influence on how the body ages. Age evaluation is much more difficult in the older generation, with a larger margin of error.

All evaluations should start with a psychological assessment to identify information that the individual can remember or perhaps incidents that they were involved in. These can help the clinical assessor to home in on an age bracket for the individual. Beyond this, we must rely on skeletal indicators. Dental age is useful in gauging a juvenile's age, but of limited help for older people. Assessment of the skeleton requires the use of X-rays (although in some cases this raises ethical problems). Either flat-plate radiography or CT images are ideal.

In older people a combination of factors helps to build up an age picture. These might include the degree of closure of the cranial sutures; evidence of degenerative conditions including osteoarthritis and the degree of ossification of cartilaginous structures, such as the costal cartilages that give the thorax its elasticity; changes to the laryngeal apparatus and the pubic symphysis; and even the extent of bone loss.

There is no single feature that will assign age with accuracy in an adult, but a multifactorial approach, combined with an understanding of the variations between populations, can provide a good estimate.

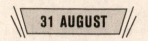

What is a blue moon?

THE TERM 'BLUE moon' comes from the traditional agricultural naming of the full moons throughout the year. The 12 full moons we see each year are named according to their relationship with the equinoxes and solstices. The names vary in different regions, but well-known examples are the harvest moon, which is the first full moon after the autumnal equinox, and the hunter's moon, which is the second full moon after the autumnal equinox. Similarly the Lenten moon, the last full moon of winter, is always in Lent, and the egg moon (or the Easter moon, or paschal moon), which is the first full moon of spring, is always in the week before Easter.

By this system there are usually three full moons between an equinox and a solstice, or vice versa. However, because the lunar cycle is slightly too short for there to always be three full moons in this stretch of time, occasionally there are four full moons. When this happens, to ensure that the full moons continue to be named correctly with respect to the solstices and equinoxes, the third of the four full moons is called a blue moon.

There are seven blue moons in every 19-year period. The next one is due to be 31 August 2022.

Genuinely blue moons can appear if volcanic eruptions or fires inject particles with a fairly uniform diameter of around a micrometre into the atmosphere. This diameter is just bigger than the wavelength of red light, which is around 650 nanometres. For example, the particles released by the 1883 eruption of Krakatoa caused the moon to appear blue for nearly two years.

SEPTEMBER

How big are mole networks?

THE TERRITORY OF an adult mole usually covers between 2,000 and 7,000 square metres, with males likely to have larger territories than females. Depending on the soil type, there may be as many as six levels of tunnel lying beneath the turf.

The depth and extent of a mole's tunnel system will vary considerably depending on a number of factors such as the type of soil and the height of the local water table. Most of the tunnels that a mole constructs are actually sophisticated traps for the numerous invertebrates on which it feeds. Earthworms and other invertebrates that enter the tunnel system are the moles' main source of food, so it is likely that a mole living in a worm-rich meadow will need a less extensive tunnel system than a mole that inhabits a tunnel system in an acidic soil where worm numbers are much lower.

Shallow tunnels are created by the mole pushing its way through the earth and bracing its body to compress the soil into walls around it. However, deeper tunnels require true excavation with the mole digging out soil into the tunnel directly behind it, and then performing a somersault and subsequently bulldozing the loose soil up to the surface to form the kind of molehill with which we are all familiar and which is a constant bane of the landscape gardener.

Building an extensive tunnel network requires a substantial investment of labour which perhaps explains why, once built, the system is fiercely defended. Maintenance of an established tunnel system requires much less effort. Territories do often overlap and, where any tunnels meet, moles leave scent signals to clearly establish their boundaries. If an owner goes absent for any length of time and those scent markings disappear, the tunnels will be taken over very quickly by rival moles.

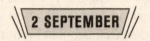
2 SEPTEMBER

How strong is a mushroom?

LOOK DOWN AT a pavement, and you might spot a toadstool growing, seemingly having pushed its way through the pavement itself.

But two inches of asphalt is nothing to the muscular mushroom. One large shaggy ink-cap (*Coprinus comatus*) discovered in Basingstoke lifted a 75 by 60 centimetre paving stone 4 centimetres above the level of the pavement in about 48 hours.

Historically, mushrooms often sprang up in foundries, supposedly from horse manure used in preparing loam for casting, and were often reported as having lifted heavy iron castings. Presumably these would have been some type of field mushroom such as *Agaricus campestris*. Whatever the species, the mechanism by which the force was exerted is likely to be the same, namely hydraulic pressure.

The upward pressure comes from the turgor pressure of the individual cells making up the wall of the hollow stalk of the mushroom. Each individual cell grows as a vertical column by inserting new cell wall material uniformly along its length.

The major structural component of the cells is a shallow helical arrangement of fibres of chitin winding round the axis of the cell. These chitin fibres are embedded in matrix materials, making the wall material like a carbon fibre composite. Chitin is an exceptionally strong biopolymer (also used by insects for their exoskeletons) and gives immense lateral strength to the fungal cell wall, so that internal pressure is confined as a vertical column. Water enters the cell by osmosis, and the resulting turgor pressure provides the vertical force that allows the mushroom to push up through the asphalt.

This phenomenon was first investigated 75 years ago by Reginald Buller, who measured the lifting power by loading weights onto a mushroom that was elongating inside a glass tube. He calculated an upwards pressure of about two-thirds of an atmosphere. The cells have a gravity-sensing mechanism that keeps the mushroom exactly vertical. A mushroom that is put on its side will rapidly reorient to grow vertically again.

As Buller found, the exquisite and fragile *Coprinus sterquilinus* exerts an upward pressure of nearly 250 grams with a stem 5 millimetres thick, so it is not surprising that more robust species can tear the tarmac.

3 SEPTEMBER

Why do we swear?

SWEARING 'RECRUITS OUR expressive faculties to the fullest', wrote Harvard psychologist Steven Pinker in his book, *The Stuff of Thought*. Yet despite being a showcase for creativity, swear words are taboo in virtually all societies, even though their subject matter – usually sex or excretion – describes activities fundamental to human existence. So why are we such potty-mouths, and what gives certain words the power to shock?

One theory is that cussing is the form of language that comes closest to a physical act of aggression. When you swear at someone, you are forcing an unpleasant thought on them and, lacking earlids, they are helpless to repel this assault. Most of us are able to restrain ourselves from launching these linguistic assaults – at least some of the time – but studies of people who lack this restraint are revealing.

Individuals with Tourette's syndrome have characteristic tics such as blinks and throat-clearing, and between 10 and 20 per cent also exhibit involuntary swearing, otherwise known as coprolalia. People with Tourette's have damage to a part of the brain called the basal ganglia – clusters of neurons buried deep in the front half of the brain that are known to inhibit inappropriate behaviour.

As Pinker sees it, the basal ganglia are responsible for tagging certain thoughts as taboo. When the 'don't-go-there' label is no longer applied, as with Tourette's, taboo thoughts can reassert themselves and the urge to cuss becomes overwhelming. There is even one recorded case of a man with Tourette's who was deaf from birth and expressed his coprolalia within his signed speech. What's more, rather than flipping the finger or making other obscene gestures that hearing people deploy, he used the recognised signs for rude words.

By poring over our rich library of filth, researchers have been able to get a handle on just what makes a good swear word. It is not just its sound: after all, 'shot', 'ship' and 'spit' are not considered obscene, whereas 'shit' is. Besides, the equivalent in French, say, sounds quite different and still packs a very satisfactory punch. It cannot just be about semantic content either, because the use of words denoting faeces or sexual matters in a medical context remains acceptable. Something about the pairing of certain meanings and sounds has a potent effect on people's emotions.

Swear words also go in and out of fashion in line with the taboos they breach. 'Poxy', 'leprous', 'canker' and other disease-related words went out of fashion as hygiene improved. Today 'fuck' reigns supreme, but there is still room for innovation. So, what will be the next big thing in swearing? Most experts decline to predict any winners, but one thing they can say is that you can't impose swear words on a language – they have to arise organically. Just because you personally dislike cheese, shouting 'stilton' out loudly is unlikely to catch on.

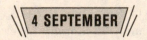

What is the point of an appendix?

ALTHOUGH IT USED to be believed that the appendix had no function and was an evolutionary relic, this is no longer thought to be true. Its greatest importance is the immunological function it provides in the developing embryo, but it continues to function even in the adult, although it's not so important and we can live without it.

The function of the appendix appears to be to expose circulating immune cells to antigens from the bacteria and other organisms living in your gut. That helps your immune system to tell friend from foe and stops it from launching damaging attacks on bacteria that happily co-exist with you.

There are other parts of the body that appear to do the same thing. Peyer's patches in the intestine help to expose your immune system to the usual contents of the intestine. By the time you are an adult, it seems your immune system has already learned to cope with the foreign substances in the gastrointestinal tract, so your appendix is no longer important. But defects in these immune sampling areas may be involved in autoimmune diseases and intestine inflammation.

Interestingly, the appendix has been used as a personal 'spare part' in surgery. It can be removed and its tissue used in reconstructive surgery of the bladder without risking the immune reaction that would be triggered by using tissue from another individual.

Is there such a thing as a negative placebo effect?

PLACEBOS ARE SUBSTANCES with no pharmacological properties, such as sugar or dummy pills. They are widely used as a control in experiments to test the effect of medicines, and are made to look and smell the same as the drug being tested. Subjects are not told whether they are receiving the actual medicine or the placebo.

How the placebo effect works is still controversial, but it is widely believed that the effect is psychological rather than physiological: the benefit occurs because people believe that the pill they are taking should cause positive effects. The effect has also been attributed to conditioning: patients expecting the effects of a drug will then experience them.

Take the example of placebos used in tests on analgesic drugs. An explanation for the placebo mechanism in this case is that it involves the release of opiate-like pain-relieving chemicals in the brain. One study found that pain was reduced by a placebo medication that the patients believed was a pain reliever, but that the effect ceased when the patients were given a drug that counteracts the effect of opiates.

The negative effects of placebos are called *nocebo* effects, nocebo being Latin for 'I will harm'. Patients receiving dummy pills sometimes experience side effects such as anxiety and depression. This is thought to be associated with the person's expectations of adverse effects of the treatment as well as conditioning. It was reported in one trial that women who believed they were prone to heart disease were nearly four times as likely to die from heart disease as women with similar risk factors who had no such belief.

Placebos pose an ethical dilemma. They work primarily by a doctor deceiving their patient into believing that they are receiving an active medicine, while in fact depriving them of any such medicine. If they also suffer nasty side effects through the nocebo effect, this arguably makes things worse.

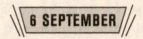

6 SEPTEMBER

How can you build the best sandcastle?

ANYONE WHO'S MADE a sandcastle before will tell you that the first thing you need is damp, rather than dry, sand. But it's not just because the water makes sand stickier.

Sand consists of lots of small, hard grains that can slide over each other. When the sand is damp, each grain is coated with a thin film of water, which tends to collect at the points where the grains touch. Surface tension acts at the surface of the water, producing the same result as if the water were covered by a stretched skin that is always in tension.

Where the water droplets adhere to the sand grains, the tension is applied to each grain and this effectively pulls each one against its neighbour, providing quite a strong force between them, even if there is no weight of sand above. The effect of this force, related to capillary action, is enough to provide plenty of friction so that your sandcastle stays together.

As you start adding water to the sand, it forms 'pendular' bridges between the sand grains which hold them together, thanks to surface tension and friction. These forces acting together can resist gravity and prevent the walls of the sand castle from collapsing. The surfaces of these liquid bridges are concave. This generates further capillary action, or suction, which also helps to hold the grains together. If you add a little more water to the sand the pendular bridges start to merge and the sand/water mix passes through the funicular state to reach the capillary state named after this force. In this state the concave liquid surfaces continue to generate a capillary action, which holds the sand grains together.

However, if you keep adding water you reach a point where the surface curvature of the liquid becomes convex rather than concave and the capillary action disappears. This is known as the droplet state. The water no longer creates any attractive force between the particles and the walls of the castle begin to slump and flow as a liquid slurry.

Obviously, you can try this out at home, but the experiment is much better if carried out on a beach, under a blue sky, with the waves lapping at your toes and a plentiful supply of ice cream.

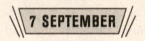

7 SEPTEMBER

Why do some foods taste better hot?

WHAT WE NORMALLY refer to as 'taste' is more correctly termed flavour, which is made up of taste, irritation and aroma. Taste per se consists only of the five sensations that can be detected by the tongue: salt, sweet, sour, bitter and umami. These are not affected by temperature and nor is irritation from, for instance, chilli peppers. But aroma, which is sensed in the nose, is strongly affected by food temperature because it depends on the release of volatile oils. The higher the temperature, the more volatiles are released, and the stronger the aroma and thus the total flavour sensation.

The flavour of foods that have little aroma is enhanced by heating, whereas foods with strong aromas may become overpowering at high temperatures. Red wines, for instance, tend to be drunk at room temperature with meals that have strong flavours, so achieving a balance in which food and drink complement each other, rather than cancelling each other out. White wines, on the other hand, are often drunk cold with fish or weakly flavoured foods.

Another important effect of temperature on meals is its influence on the viscosity of starch-thickened sauces, which drops at higher temperatures because starches react to heat. The texture of food is very important to people. A meal covered in a cold, starch-thickened sauce is pretty unappealing, while a non-starch-thickened sauce such as mayonnaise covering the same ingredients in a sandwich would be a very different prospect.

There is also a large element of convention and cultural preference involved. We prefer our gazpacho cold but our minestrone piping hot. Beer is served at room temperature in the UK but chilled almost everywhere else. Some people prefer whisky on the rocks, others – especially in Scotland – find ice an abomination. Hot coffee and iced coffee are equally acceptable to most people, and choice depends mainly on ambient temperature. It's all about circumstance, accompanying flavours, and how we are used to having our food and drink served.

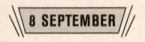

8 SEPTEMBER

How do black boxes work?

FLIGHT DATA USED to be recorded on photographic film that had to be housed in a box into which light could not penetrate. This is one explanation of the origin of the term 'black box' recorder.

The black box now comprises the flight-data recorder and the cockpit voice recorder. There are also calls for the addition of a cockpit image recorder, which would record the external readings of the instruments and therefore what the flight crew actually sees.

David Warren was the first to develop a prototype of a combined data and voice recorder in 1957. As a research scientist at the Aeronautical Research Laboratory (ARL) in Melbourne, Australia, he helped to investigate a series of fatal accidents involving the De Havilland DH106 Comet in 1953 and 1954. He recognised that access to a recording of what had happened in the aircraft before the crashes would have been invaluable.

The aviation community gave the black box a largely lukewarm reception at first, until the crash of a Fokker Friendship at Mackay in Queensland, Australia, in 1960. This prompted Australia to make the black box recorder compulsory, and other aviation authorities followed suit.

According to the Aviation Safety Network, about 2,300 commercial airliners have suffered a breach of their fuselage, known as a hull breach, since then. As well as crashes and mid-air collisions, airliners have been shot down, or been the target of terrorist bombs and hijackings. The average number of hull breaches is now about 30 a year.

Of course there are instances where the cause of a crash is still uncertain even when the black box is recovered, but investigators have failed to recover the black box in only ten hull-breach incidents, which equates to less than 0.5 per cent of crashes.

As well as recording flight data to a black box, limited data is transmitted. However, communication with ground stations is sometimes lost, and even transmission to satellites is not perfect. Even when encrypted, there is always the worry that transmitted data could be hacked, redacted or lost before it reaches accident investigators.

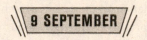

Where are the largest tidal ranges found?

Tidal range is the height difference between high tide and low tide. The largest tidal ranges in the world – of up to 16 metres – are found in the Bay of Fundy, on Canada's Atlantic coast. To put that into context, the typical tidal range in the open ocean is about 0.6 metres.

To understand what happens in the Bay of Fundy, start with a hand basin half-full of water. Push down on the surface on one side with the palm of your hand and the water will rise on the other, after which it will slosh back and forth like a liquid see-saw. By pushing down even more on each side at the same time as the level on that side is falling, the rise and fall of the surface will increase, and can be made to overflow the rim of the basin. The sloshing of the water has a natural frequency and your additional input resonates with it, increasing the amplitude of the see-saw wave. At the central axis of the basin the level remains unchanged, although water moves to and fro horizontally.

Now imagine the basin cut vertically down that central axis and consider just half of it. The half that is left corresponds to the Bay of Fundy, the axis-edge marks the opening of the bay, and the missing half of the basin is replaced by the open Atlantic Ocean. An incoming tide effectively appears in the ocean as a huge wave advancing towards the bay. As it reaches the continental shelf at the bay's opening it plays the role of the high half of the see-saw and happens to coincide with the low water level at the far end of the bay. By the time this wave has moved to the innermost section of the bay – raising its water level to a peak – the dip in the ocean surface corresponding to the low tide has reached the continental shelf.

The exceptionally high tides occur because the successive incoming tides appear at nearly the same frequency as water sloshing into and out of the bay, just as happened in the hand basin. It is a resonance effect.

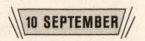

Why is it harder to lose weight as you age?

THERE ARE A number of contributing factors at work here. As we age we tend to reduce our active pursuits and, above the age of 20, our resting metabolic rate reduces by about 3 per cent per decade, allowing our bodies to go further and further on the same amount of food. On top of this, as we age, we use less energy to digest and process the food that we eat.

There are also hormonal changes that encourage fat build-up. A decrease in levels of growth hormone and testosterone contribute to an increase in fat mass. Additionally there is a reduced response to hormones produced in the thyroid gland, which stimulate metabolism of lipids and carbohydrates, and possibly also to leptin, which helps to control appetite.

One more factor is our gradual reduction in height, and a reduction in the weight of organs and muscle mass. For example, neurologists Anatole Dekaban and Doris Sadowsky identified a progressive fall in brain weight from about 45 years of age onwards.

Most of us do not compensate for these factors by reducing energy intake, so we gain weight. Given this combination of factors, it is easy to understand why we find it more difficult to get rid of our excess weight as we age: our bodies are conspiring against us.

However, if we reduce our total energy intake while increasing our activity – especially activity that builds muscle mass – we have a better chance of losing, or at least maintaining, weight.

Why do leaves change colour?

THE SEASON OF autumn is characterised by golden trees, and crispy leaves crackling underfoot. But apart from looking pretty, why do leaves go through such colourful changes throughout the year?

In autumn months, chlorophyll is broken down so that its components can be recycled by the plant. The extremities of the leaf are exposed to the greatest variations in temperature, so chlorophyll breaks down there first. This reveals the underlying yellow carotenoid leaf pigments, hence the edge of the leaf is yellow with splashes of orange-brown as the leaf cells finally die. This region will spread inwards until the whole leaf is brown or the leaf falls off the tree.

Over the major veins there is only a very thin layer of photosynthetic cells, which means their chlorophyll is broken down early compared with other parts of the leaf. In contrast, the green stripes that appear between the veins are areas where the leaf is thickest and the photosynthetic cells are protected for longer from the changes in environment. In this case, the layers of cells near the leaf surface have probably already turned yellow but this is masked by the cells below that still contain chlorophyll. The green patches look especially dark but this is probably an optical illusion resulting from the surrounding yellow tissue.

No two leaves are in identical condition as they die in autumn, because of the differing microclimates experienced by each leaf, and thus the development of pigment patterns is remarkably variable. Never expect to see the same pattern twice.

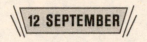

12 SEPTEMBER

Why do some people attract mosquitoes?

WHILE IT IS true we all produce a bouquet of natural chemicals, some of which attract biting insects, a fortunate few give off an aroma that apparently masks these attractive chemicals, and so prevents these people from being bitten because the mozzies can't track them down. James Logan and John Pickett at Rothamsted Research in Hertfordshire, UK, have pinpointed the human chemicals that keep blood-sucking insects at bay and are developing a mosquito repellent based on these chemicals.

The team at Rothamsted first noticed that some individuals are less appealing to blood-sucking insects when they saw some herds of cattle having to fend off flies while others grazed undisturbed. Pickett's group reasoned that animals in the undisturbed herd must simply smell less alluring to the flies, and examined the cows' chemical profiles. Sure enough, they discovered that very distinctive chemical signals emanated from certain individuals, and these lucky cows were unattractive to flies.

Logan discovered that humans have distinct chemical signatures too. He let yellow fever mosquitoes (*Aedes aegypti*) fly along a Y-shaped maze and wafted the scent from volunteers' hands down the prongs of the maze. While some chemicals attracted mosquitoes, others repelled them, and so the insects flew down the more attractive prong of the Y. Logan has since been able to identify which chemicals the insects respond to by strapping miniature electrodes to the antennae of female mosquitoes and checking their responses. It turns out that only some humans produce the masking chemicals that mosquitoes find unappealing.

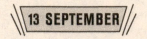

Why don't penguins' feet freeze?

PENGUINS, LIKE OTHER birds that live in a cold climate, have adaptations to avoid losing too much heat and to preserve a central body temperature of about 40 °C. The feet pose particular problems since they cannot be covered with insulation in the form of feathers or blubber, yet have a big surface area (similar considerations apply to cold-climate mammals such as polar bears).

Two mechanisms are at work. First, the penguin can control the rate of blood flow to the feet by varying the diameter of arterial vessels supplying the blood. In cold conditions the flow is reduced, when it is warm the flow increases. Humans can do this too, which is why our hands and feet become white when we are cold and pink when warm. Control is very sophisticated and involves the hypothalamus and various nervous and hormonal systems.

However, penguins also have 'counter-current heat exchangers' at the top of the legs. Arteries supplying warm blood to the feet break up into many small vessels that are closely allied to similar numbers of venous vessels bringing cold blood back from the feet. Heat flows from the warm blood to the cold blood, so little of it is carried down the feet.

In the winter, penguin feet are held a degree or two above freezing to minimise heat loss, whilst avoiding frostbite. Ducks and geese have similar arrangements in their feet, but if they are held indoors for weeks in warm conditions, and then released onto snow and ice, their feet may freeze to the ground, because their physiology has adapted to the warmth and this causes the blood flow to feet to be virtually cut off and their foot temperature falls below freezing.

If you travelled at the speed of light, would there be a flash like a sonic boom?

THERE IS AN analogue to the sonic boom for electromagnetic waves, but the particle must be travelling faster than the speed of light. This is possible because light travelling through a medium has a velocity less than its velocity in a vacuum – nothing can travel faster than light in a vacuum. This lower velocity is given by $v = c/n$ where c is the speed of light in a vacuum, and n is the refractive index of the medium.

The radiation of light analogous to the sonic boom, called Cherenkov radiation, is produced whenever the velocity of a particle exceeds c/n. The blue glow that emanates from water in which highly radioactive nuclear reactor fuel rods are stored is caused by the Cherenkov effect. Much of the radiation that fuel rods emit is in the form of high-energy electrons. The electrons travel through the water at a velocity greater than that of light in water and hence cause the characteristic 'Cherenkov glow'.

The importance of the Cherenkov effect as a scientific tool lies in the connection between a particle's momentum and the angle at which the Cherenkov photons are emitted. A measurement of the angle of Cherenkov emission provides an indirect measurement of the speed and direction of a particle. Cherenkov detectors are one of the important tools used by particle physicists to probe the ultimate small-scale structure of matter.

Is there a continuum of consciousness?

IT APPEALS TO common sense that as animals become more complex they also get ever more consciously aware. For one thing, it's gratifying to have the continuum end with us. But common sense is often hopeless at explaining anything. Take consciousness itself. It's a word describing a phenomenon that everyone understands, but that is incredibly difficult to pin down scientifically.

One of the leading theories of consciousness, integrated information theory (IIT), sidesteps this problem by proposing that the experience of consciousness is constructed by data drawn from various locations in the brain. IIT says there is a measurement, phi, which represents the amount of emergent information possessed by a system. If IIT is right, then it implies that all animals are to some extent conscious, and that there is indeed a continuum moving from simple to complex animals. But because we don't know where to look or how to measure it, it's hard to say where any of them are on the continuum. Or indeed, where we are.

We have no reason to think that human consciousness is the 'most' conscious a being can get. For a start, other animals have forms of consciousness without the same brains as us. Birds don't have a neocortex, the part of the brain we used to believe, with mammalian superiority, was essential for conscious thought. Instead they manage to do lots of complex cognitive processing – thinking – using a different part of the brain. Most biologists who have worked with great apes will assure you they are conscious, too. You can even make the argument for plants. They are remarkably good at detecting and processing huge amounts of information, and acting on it. That's certainly something we'd call awareness, and maybe even consciousness.

So it's not really controversial to state that there's a continuum of consciousness. Each of us also experiences the points along the line. In our dreams we are still mostly ourselves, even if we happen to be flying or breathing underwater. If we're drunk or on drugs our consciousness operates differently. As IIT originator Giulio Tononi of the University of Wisconsin, Madison, puts it, consciousness was different when we were very young and will be different again when we are very old.

We're certainly not at the peak of conscious possibility. What's next? Telepathy? Prophesy? Possibly a higher consciousness would be far more attuned than we are to sensing another being's conscious state, and predicting its behaviour. To us this extreme empathy or hyper-intuition would look like mind-reading but would not require supernatural explanation, only a deeper understanding of neural processing.

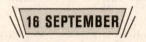

16 SEPTEMBER

Why does turmeric stain so much?

TURMERIC, THE POWDERED rhizome of *Curcuma longa*, and paprika, which is obtained from the fruits of sweet peppers, *Capsicum annuum*, are examples of spices used in cooking as much for their colour as for flavour.

The yellow colour of turmeric is caused by curcumin, which makes up around 5 per cent of the dry powder. The red pigments in paprika are a mixture of carotenoids, principally capsanthin and capsorubin, and in dried paprika they amount to a maximum of 0.5 per cent of the weight.

The red carotenoids, which consist of long, chain-like molecules, are soluble in organic solvents such as petroleum spirit. Curcumin consists of smaller molecules with terminating phenyl groups. It is insoluble in water but dissolves in solvents like methanol. Therefore you might expect that both paprika and turmeric would stain paintwork and plastics, because they dissolve in organic solvents. You would also expect them to migrate to the oily part of food during cooking.

To compare their colouring properties, place a good pinch of turmeric into two small glass spice jars and do the same with paprika. Add a dessertspoon of methylated spirit to one set and the same amount of white spirit to the other (you can repeat the experiment with cinnamon and chilli powder). Upon shaking the mixtures you will see a vivid yellow colour appear instantly in the meths from the turmeric powder and the white spirit turn red from the paprika. When you place a drop of extract from each of the four jars on to a clean white plate, you will see that the turmeric-meths extract has much the strongest colour followed by the white spirit and paprika. The same experiment can be done with acetone (nail varnish remover).

This demonstrates the principal reason why turmeric stains more than other spices – it simply has more extractable colouring material in it. Other reasons will reflect the different physical properties of curcumin and the red carotenoids, as demonstrated in our solubility experiment, and differences in the way the dyes react chemically with solid materials.

Curcumin is stable when heated but is not stable when exposed to light. So to remove a turmeric stain, first clean with methylated spirit and then place the object in sunlight.

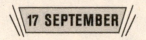

Do trees inherit genetic traits?

PEOPLE OFTEN SAY a child has its mother's or father's eyes. It turns out that trees can inherit visible genetic traits in much the same way.

The characteristics found in trees are under varying degrees of genetic and environmental control, just like the traits of other organisms. For a given characteristic, the proportion of total variation that is explained by genetic control is known as its heritability. This can range from 0 to 100 per cent: the higher the value, the more closely progeny resemble their parents and the greater the improvement that can be obtained by selective breeding.

Two major sets of visible traits are under a high degree of genetic control and are of great importance for the end use of the wood. One is 'stem form', which is a measure of sinuosity or deviation from perfect straightness. It is associated with 'reaction wood', which forms in response to mechanical stress, such as exposure to strong wind. Reaction wood is visibly asymmetric and is generally undesirable for end uses such as solid wood and pulping for paper.

The variable characteristics of the tree's branches are also important to industry and include the thickness, number per unit of stem length, angle of insertion on the stem, and whether branches occur in whorls or are scattered at random along the length of the stem. These variables affect the number and size of undesirable knots.

Other characteristics in which parents and progeny may be similar include flower colour in horticultural species and flower or fruit colour, shape, size, flavour and nutrient content in fruit trees.

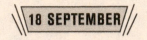

Why do hens cackle after laying an egg?

FOR A HEN, there are many steps that go into the routine of laying an egg, and they can usually be seen performing similar actions day to day.

A hen will often enter the nest ten to twenty minutes before laying and settle herself. She stands to lay the egg which is soft-shelled and hardens on contact with the air. She then sits quietly for a few minutes, preening, crooning and resting. The hen then jumps up from the nest with a loud cackle and given the freedom of the range, runs a considerable distance. A nearby dominant rooster hears the cackling, runs with wings outstretched to the hen, and mates with her immediately. He then performs a 'stamping' display and both go off happy.

It seems that the post-egg cackle helps to attract the rooster to mate. If there are rival males, the rooster may patrol outside the nest like any expectant father and then mate with the hen as soon as she has finished laying. In that case, the cackle may be quiet to non-existent. When mating at other times, the stamping display comes beforehand and there is no cackle.

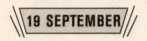

Why do wet things smell worse than dry things?

MAKING SOMETHING WET does not automatically make it more smelly. For instance, a wet clean towel smells no worse than a dry clean towel.

However, the presence of moisture does allow the growth of bacteria, assuming that there is organic matter present for the bacteria to eat. As they grow and multiply, bacteria produce a whole range of smelly compounds of the kind you can detect in bad breath, for example. Most are fatty acids, amino compounds and the like, with charged chemical groups that readily bind to non-volatile molecules such as large proteins and carbohydrates. Once they have latched onto, say, dry cloth or leather, they cannot float freely into the air so there is not much to smell. However, these charged groups have an affinity for polar molecules, and the most polar of common molecules is water. So when the object gets wet, water molecules prise loose the odour molecules, cocooning them in tiny mobile parcels of water. For good or ill many escape into the air, reaching nearby noses in vast numbers.

So, given moisture and enough time for bacterial growth, wet things can smell worse. But if you prevented bacterial growth by, say, sterilising the wet item to kill all bacteria, then it wouldn't develop such a smell. Another option is to release other molecules that immobilise pong molecules by binding them with complementary charged groups. Chlorophyll combats smells partly by presenting a metal atom that binds the active groups of many smell molecules. Similarly, by binding key molecules, partly oxidised paraffin wax vapour from the smoke of burning candles also helps clear a room of the stench of cigarettes.

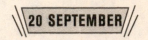

How long would it take a coconut to float from the Caribbean to the west coast of Scotland?

THE COCONUT PALM seed (*Cocos nucifera*) is the best known of the drift fruits, and it is claimed that viable coconuts have been found as far north as Norway. However, these may have been tossed from ships into the North Sea rather than drifting all the way from the Caribbean. The chances are that a coconut would sink long before reaching Scotland, despite being carried by that 'river in the ocean', the Gulf Stream.

There is more chance of finding flotsam lost from the cargo vessels that ply our oceans, however. For example, in 1992 an armada of 29,000 rubber ducks and other bath toys were spilled overboard during a storm from a container ship as it crossed the Pacific Ocean. Curtis Ebbesmeyer, a retired oceanographer, has been tracking their progress. Now bleached white but still identifiable because of the logo 'The First Years' that is stamped on them, each duck has a $100 price on its head, an incentive for beachcombers to report their finds and help scientists develop better models of our oceans.

It is thought that a flotilla of these ducks has reached the Atlantic by navigating their way through the Northwest Passage. The ducks have already proved that flotsam travels up to twice the speed of ocean currents.

Using this observation and knowing the varying speeds of the Gulf Stream and North Atlantic Drift, it would take the hypothetical coconut about 16 months to make the journey from the Caribbean.

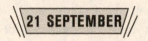

Why does hair turn grey?

GREY (OR WHITE) is merely the base 'colour' of hair. Pigment cells located at the base of each hair follicle produce the natural dominant colour of our youth. However, as a person grows older and reaches middle age, more and more of these pigment cells die and colour is lost from individual hairs. The result is that a person's hair gradually begins to show more and more grey.

The whole process may take between 10 and 20 years – rarely does a person's entire collection of individual hairs (which, depending on hair loss, can number in the hundreds of thousands) go grey overnight. Interestingly, the colour-enhancing cells often speed up pigment production as we age, so hair sometimes darkens temporarily before the pigment cells die.

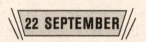

Why do some clothes become static?

STATIC ELECTRICITY IS an imbalance of electric charge: a lack or overabundance of electrons on the surface of the material. This typically occurs by 'tribocharging' when two materials are brought into contact then separated, electrons are exchanged by the materials, leaving one with a positive charge and the other with a negative charge. Friction between the two materials can enhance this charge-separation process.

The amount of static build-up is also highly dependent on

the relative humidity – the higher the humidity the lower the charge. Fibres such as rayon, silk, wool, cotton and linen have high moisture 'regain' – their fibres absorb a great deal of moisture at a given humidity from a bone dry condition – and are low in static. Fibres such as polyester, acrylic and polypropylene, having low moisture regain, are high in static.

A layer of fabric conditioner can reduce the electrical resistance of the surface of fabrics.

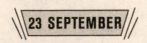

23 SEPTEMBER

What would an alien look like?

THE FAMOUS GENETICIST Conrad Waddington believed that any higher life form would have to look like . . . Conrad Waddington. But most people see evolution as a contingent process – in other words, if evolution on Earth was run through again, land vertebrates – and that includes us – would be unlikely to reappear. And if they did we'd look very different. Of course, this applies to other planets too.

So, if we can't have humans on other planets, what can we have? There are patterns of general problems, and common solutions, that apply to life anywhere in the universe. We know this because different species on Earth invent identical solutions separately. Birds, bats, insects and some fish all fly. And plants and some bacteria photosynthesise. These universal solutions will be found on pretty well all other planets with life – including intelligence. So life will be formed of universal solutions – such as the elephant's huge legs to support great bulk in gravity – and local or parochial ones – such as its trunk, which developed from a need, on Earth, to pick up food from its feeding spots. Its food on another planet might not have required trunk-to-mouth delivery.

The difficulty, therefore, is in recognising universal solutions – which aliens will possess – and parochial ones – which they will not. Parochials normally happen only once – the trunk – universals more times – flying. Joints seem universal; the number of digits on a limb is not. Eyes, yes, external ears, probably not. And the list is very diverse . . . our strange excitement in sexual guilt pleasures is almost certainly parochial, so alien pornography seems unlikely.

This means the standard clichés of science fiction would not hold up. *Star Trek*'s Mr Spock, whose anthropomorphic appearance and evolutionary convergence is so close to humans that he can inter-breed with us when we cannot even breed with species on our own planet, is, sadly, illogical. And we should disbelieve all the flying saucer stories that have little green men, not because they are little and green, but because they are men. Little green splots are so much more believable.

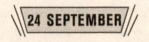

24 SEPTEMBER

Can you wash your clothes with conkers?

NOT ALL DETERGENTS come in packets; some grow on trees. Take horse chestnuts, for instance. When you're not threading them onto shoelaces, baking them in the oven or soaking them in vinegar to win conker championships, you could be using them to wash your whites whiter.

To do this, grab some conkers, remove the brown outer casing, chop them up into small pieces, and put them in a pan. Add a cup or two of water and boil them for a few minutes, then let them cool. Strain the mix through a tea towel into a washing-up bowl to remove the solids and keep the liquid. Pour this into a bottle and shake it. Now put it back in the bowl and wash your socks in it.

You should see a soapy lather form on the liquid you pour into the bowl. And, if you give your socks a good scrub, you'll see how clean they emerge. This is because horse chestnuts contain a saponin, a natural soap or surfactant. As we have seen, it can be extracted with water – a trick that has been used for centuries to make a soapy liquid for cleaning linen.

Surfactant molecules have a polar region that is attracted to water molecules (it is hydrophilic) and a non-polar region that is repelled by water (hydrophobic). They are therefore soluble in both water and organic solvents, including the substances that make your socks dirty. This makes them very similar to synthetic detergents and very good for cleaning. As they are so mild, they are also popular with art conservators, who use them to clean delicate fabrics or ancient manuscripts.

Be aware that the conker extract you have created is mildly poisonous if drunk, causing coughing and sneezing, and can be a skin irritant. So, close adult supervision of children is required, and when you've finished, give all the utensils you used a good wash to make sure there are no traces of saponin left.

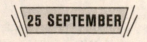

What causes cells to stick together in the human body?

CELLS IN THE body are organised in tissues that are held together through a variety of molecular interactions.

On the one hand, cells interact with each other. This is a very specific interaction mediated by various families of adhesion molecules called cadherins, neural cell adhesion molecules and intercellular cell adhesion molecules. These are all expressed on the surfaces of cells and anchored in the cytoskeleton of each cell, an arrangement which stabilises and gives strength to interactions between cells.

On the other hand, the body's tissues are not made up solely of cells, but of an intricate network of macromolecules too, called the extracellular matrix. These macromolecules are combined into an organised mesh and, depending on the proportions of its components, the matrix can adopt diverse forms adapted to particular functional requirements. For example, it can be calcified and hard as in the bones and teeth, transparent as in the cornea, or elastic and strong as in the tendons. The main components of the matrix, which determine the properties listed above, are fibre-forming proteins that can be structural (collagens and elastin) or adhesive (fibronectin and laminin).

Cells adhere to this complex scaffolding through surface receptors called integrins that are anchored in the cell cytoskeleton and bind to the matrix components. Though integrins are densely packed on the cell surface, they have a relatively low affinity for interacting with the matrix components. This allows the cells to move within the matrix without losing their grip completely, meaning that it is, in effect, a rather flexible glue.

However, the interactions between integrins and matrix components have a deeper purpose than just holding the cells in place. Almost like antennae, they can transmit messages to the cell about the microenvironment to which it needs to adapt, and so influence cell shape, movement and function.

There are also, of course, cells in the body that remain free. These are the components of the blood: red and white blood cells and platelets that normally float in the bloodstream, delivering oxygen to the tissues and keeping a look out for invading micro-organisms and wounds. For example, the encounter of a platelet with a wound activates the integrins of the platelet, enabling it to bind to ibrinogen in the blood vessel and initiate the aggregation process that forms clots and stems bleeding.

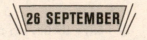

If all boats were taken out of the sea, how much would sea level drop?

ACCORDING TO ARCHIMEDES' principle, a body sinks until it displaces its own weight of water. This is why a fully laden ship settles deeper in the water. Each tonne of ship or its cargo displaces a cubic metre of water.

Warships are referred to by their displacement. The combined displacement of the world's military fleets is about 7 million tonnes. The merchant fleet is much larger, but to complicate matters, merchant ships are referred to in terms of deadweight tonnes (dwt). This is the mass of cargo that can be loaded onto an empty ship before it risks capsizing. It takes no account of the water displaced by the ship before the cargo is stowed. The *Knock Nevis*, launched as the *Seawise Giant* in 1979 and scrapped in 2010, possessed the largest deadweight tonnage ever recorded. It could carry a cargo of up to 564,763 dwt and displaced 83,192 tonnes of water when empty.

The world merchant fleet amounts to 880 million dwt. If the empty fleet displaced the same tonnage of water as the cargo it can carry, then the fully laden fleet would displace 1,760 million tonnes. Add the displacement of the military fleet and the world's large ships will displace 1,767 million cubic metres of water. When spread over the surface of the oceans (about 360 × 1,012 square metres), sea level would rise by a mere 5 micrometres. The slightly higher density of sea water makes little difference to this figure.

Sea freight increases at a rate of about 3 per cent annually, equivalent to emptying the contents of 10,000 Olympic-sized swimming pools into the oceans. But this is, indeed, a drop in the ocean – 25,000 times smaller than the annual rise in sea level.

Are there any green mammals?

THERE IS ONLY one green mammal, the three-toed sloth.

Sloths are tree-dwelling mammals from Central and South America that are known for being very slow and sleeping a lot. Their greenness is a result of a coat of algae that covers their fur. The algae is thickest around their head and neck – where the fur is longest. Because sloths clean themselves with their hands and not their tongues, the algae-coating is never cleaned off. Isn't this very annoying for the sloth?

Actually, it's very useful. One of the sloth's main predators is the eagle. There's no way that a sloth could move quicker than an eagle, but a green sloth that moves very slowly through the trees is very tricky to spot. So the three-toed sloth isn't *really* green. In fact, there are no known mammals capable of producing their own green skin pigment.

The main reason for the absence of green mammals seems to be an ecological one. In general, mammals are simply too big to use a single colour for camouflage because there are no blocks of green large enough to conceal them. Most mammals have an environment that is made up of patches of light and dark and composed of many different colours. This means that those mammals that are camouflaged tend to be dappled or striped. Animals that do use green coloration for camouflage, such as frogs and lizards, are small enough to use solid blocks of green – leaves and foliage – for cover.

The main predators of most mammals are other mammals, especially the carnivores, such as the cat, dog and weasel families. Carnivores are all colour-blind or, at best, have very limited colour vision. Hence effective camouflage against them is not a matter of coloration but of a combination of factors such as brightness, texture, pattern and movement.

Why does asparagus make your wee smell?

THIS REALLY IS quite a surprising effect, the onset of which can occur within minutes of eating asparagus. Once you recognise the smell it's impossible to visit a public lavatory without knowing if someone has been chomping on this expensive seasonal veg.

The cause of the odour has perplexed biologists for a long time. While sulphurous compounds are known to be implicated, recent studies seem to suggest that it is in fact a cocktail effect, which combines to produce the unmistakable whiff. Methanethiol, dimethyl disulphide and dimethyl sulphone are the likely candidates. These substances are probably the result of the body breaking down the S-methylmethionine and asparagusic acid found in asparagus.

The difficulty in determining exactly which substances are responsible is compounded by the many variations in human production and perception of the odour. Several studies have indicated that production of the odour is a genetically determined trait exhibited by a maximum of 50 per cent of adults.

However, at least one report suggests that while all people produce the odour, only a minority are genetically able to smell it. So it may be that different individuals produce a different array of compounds and also have a differential sense of smell to these.

The reason that urine begins to smell so quickly after consuming asparagus may be due to the rapid response of the kidneys to any unusual products, such as the sulphurous compounds mentioned above. The body will quickly process and remove anything it considers potentially alien and the asparagus odour can often be detected in urine within 15 minutes of consuming the vegetable. Not that anybody should be alarmed – asparagus is an excellent, healthy food, it's just that your body removes the compounds it has no use for as quickly as possible while simultaneously absorbing the beneficial ones.

Not everyone finds the smell offensive. Juvenal Urbino, in Gabriel García Márquez's *Love in the Time of Cholera*, 'enjoyed the immediate pleasure of smelling a secret garden in his urine that had been purified by lukewarm asparagus'.

Are there any viruses that can make you live longer?

PLENTY OF VIRUSES will shorten your life or kill you. But are there any that make you live longer?

It is sometimes in the interest of viruses to keep their host alive long enough for the virus to make use of its cellular mechanisms to reproduce itself and spread itself through contact between the current host and other potential hosts. For this reason the myxomatosis virus rapidly became less virulent when intentionally released into the Australian rabbit population. Not only were the rabbits evolving an immunity, but natural selection favoured viruses that did not kill the host so rapidly, or even left it alive.

Some viruses are even being used in the fight against cancer. For example, researchers from the Hebrew University of Jerusalem are targeting brain tumours using a variant of the Newcastle disease virus, which usually afflicts birds.

Likewise, genetic engineers use harmless viruses to carry desirable genes into cells. For example, scientists have tried treating familial hypercholesterolaemia, a genetic disease, by infecting liver cells with a virus. The virus inserts a crucial gene that makes the liver cells produce a chemical sponge that controls harmful cholesterol.

Trying to hijack viruses for our own ends goes back at least as far as the close of the 18th century, when Edward Jenner discovered that milkmaids infected with relatively mild cowpox were immune to virulent smallpox. This eventually led to widespread vaccination using the closely related vaccinia virus.

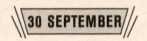

What causes a gust of wind?

NEAR THE SURFACE of Earth, friction slows the wind. Turbulence is almost always created by layers of air moving at different velocities and this enhances or reduces the surface wind. Strong turbulence is also created by obstructions such as buildings, which is why city centres are notoriously gusty.

If the surface is sufficiently warmer than the air above, then convection will produce columns or walls of warm air called thermals. These will rise from the surface, and draw in currents of air to the base of the rising column. These currents can add to the mean wind to produce gusts that are longer lived than the usual turbulent gust.

In addition, if the convection is strong enough, it may produce shower clouds by condensation of moisture in the thermal as it rises and cools. Subsequent evaporation can then result in columns of cold air rapidly descending from these clouds to produce violent gusts at the surface. These are sometimes called squalls.

OCTOBER

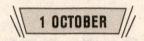

1 OCTOBER

Why does helium increase the pitch of your voice?

SOUND TRAVELS FASTER in helium than in air because helium atoms (atomic mass 4) are lighter than nitrogen and oxygen molecules (molecular mass 14 and 16 respectively).

In the voice, as in all wind instruments, the sound is produced as a standing wave in a column of gas, normally air. A sound wave's frequency multiplied by its wavelength is equal to the speed of sound. The wavelength is fixed by the shape of the mouth, nose and throat so, if the speed of sound increases, the frequency must do the same. Once sound leaves the mouth its frequency is fixed, so the sound arrives to you at the same pitch as it left the speaker. Imagine a rollercoaster ride. The car speeds up and slows down as it goes around the track, but all cars follow exactly the same pattern. If one sets out every 30 seconds, they will reach the end at the same rate, whatever happens in between.

In stringed instruments, the pitch depends on the length, thickness and tension of the string, so the instrument is unaffected by the composition of the air. Releasing helium in the middle of an orchestra would therefore create havoc. The wind and brass would rise in pitch, while the pitch of the strings and percussion would remain more or less the same. In *The Song of the White Horse* by David Bedford, the lead soprano is required to breathe in helium to reach the extremely high top note.

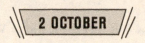
If a thermometer was in space, what would it read?

A MERCURY THERMOMETER would stop working at 234 kelvin (-39 °C) – the temperature at which mercury freezes – but there are types of thermometer that would work.

In the vacuum of space, there are virtually no particles to conduct or convect heat, so a thermometer would not be affected by these processes. However, it would still be affected by radiation from stars and cosmic microwave background radiation.

If the thermometer was in Earth's shadow and there was no moonlight, then it would read about 3 K. This would be because of cosmic microwave background radiation at 2.7 K and some infrared radiation from the night side of Earth. If it was in direct sunlight, the temperature would be about 500 K (230 °C), which is why some spacecraft such as the Apollo lunar modules were wrapped in gold foil to stop them overheating.

So the short answer is it would record anything from 2.7 K to millions of degrees, depending on where it is in space.

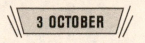

3 OCTOBER

Do elephants sneeze?

SNEEZING IS AN involuntary response that serves to remove foreign or excess material from the nasal passages. Elephants are just as liable to experience foreign matter in their nasal passages as other mammals and presumably sneeze for the same reason as dogs, cats and humans.

A story from a *New Scientist* reader, John Walters, illustrates the humorous consequences of an elephant sneeze in action:

> *I frequently camp in the bush close to where I live in northern Botswana. By far the most pleasant way to experience the African night is to sleep under a mosquito net rather than in a tent, though you may end up, as I have, being investigated by lions, hyenas, hippos and elephants, which can be quite exciting.*
>
> *A friend told me of a time when, sleeping under a net, he woke in the middle of the night and, not being able to see the stars, believed that it had clouded over and might rain. But as his eyes focused more clearly he realised that he was looking up at the underside of an elephant. Being inquisitive, the elephant was sniffing him through the net. Then, suddenly, there was an eruption from the elephant's trunk and my friend's face was covered in elephant mucus! The animal then carefully stepped over the net and went on its way.*

So, yes, elephants do sneeze. Just try not to be close to one when it does.

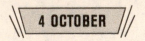

4 OCTOBER

Why are maps inaccurate depictions of the Earth?

THE PROBLEM WITH maps is that you can't peel the surface off a sphere and lay it flat without distorting it. Imagine trying to do so with orange peel. You have to choose how to distort it: you can preserve area, distance or direction, but not all of them. This compromise is inherent in all map projections of the earth's surface.

There is no perfect depiction, and choosing the appropriate one depends on the map's purpose. The sixteenth-century Mercator projection preserves compass directions: the north–south and east–west lines are straight (although all other straight journeys look curved, as you may have seen on in-flight aircraft route maps). This preservation of compass direction, as well as the accurate depiction of coastal features, is what sailors cared about – and they were the most important customers for maps at the time. But the Mercator projection has a huge drawback. It makes a 40-kilometre circle around the North Pole as wide as the 40,000-kilometre equator. Africa looks smaller than Greenland when it is actually fourteen times larger, and India looks tiny.

So an 1855 alternative called the Gall projection was revived in 1973 as the Peters projection. This squashes the vertical distance near the poles to make up for the inherent horizontal expansion. The result is that northern countries (and Australia) are unrecognisable, but the relative area of each country is conserved. If you are prepared to give up on rectangular maps, a semi-oval is a good compromise, although things at the edges still suffer: the 1805 Mollweide map is a popular equal-area projection, and the National Geographic Society uses the 1921 Winkel tripel projection.

Online maps such as Google Maps still use a simplified Mercator projection because preserving compass directions on a rectangle is their main purpose.

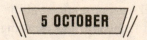

Is there any difference between a shaken vodka martini and a stirred one?

JAMES BOND IS famously fastidious in demanding that his martinis are only ever served shaken, not stirred. Apparently, there *is* a difference . . .

Mix 140ml of vodka, eight drops of vermouth, two olives and a twist of lime or lemon in a jug (gently, to avoid any undue shaking or stirring). Carefully divide the mixture into two cocktail shakers. Shake one, stir the other, then pour the martini into two separate glasses. Ideally you'll need to recruit a volunteer for a blind tasting – but don't tell them which is which. Supposedly, when a martini is shaken not stirred it 'bruises' the vodka spirit in the drink. To seasoned martini drinkers this alters the taste.

However, there is great debate as to what is really happening. It is impossible to bruise an alcoholic spirit in the way that you can bruise fruit because it is a liquid and has no vascular system, although the olive and citrus fruit (if you add them to the cocktail mix before shaking or stirring as our recipe does) can be bruised by the shaking action, thus releasing flavoursome oils and juice. Other recipes, however, eschew adding olives and fruit beforehand so that only the vodka and vermouth are shaken or stirred, thus ruling out botanical intervention in the flavour. We suggest you repeat this experiment adding the olives and fruit after shaking or stirring and see if you can tell the difference.

Nonetheless, martini drinkers can still detect a difference when the vodka and vermouth are shaken or stirred without the olives and fruit. This is because a martini is usually drunk within seconds of preparation, rather than minutes. The tiny bubbles caused by the shaking mean that a well-shaken martini is cloudy. This has an effect on the texture of the drink – it is less oily than the stirred version, and hence the taste is ever so slightly altered.

The bubbles from shaking can also partially oxidise the aldehydes in the vermouth in the way that the flavour of red wine alters when we oxidise it – commonly known as letting the wine 'breathe'. This can alter the flavour of the martini – again, ever so slightly.

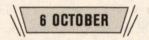

6 OCTOBER

Why do you get 'sleepies'?

'SLEEPIES', 'SAND' OR 'sleep', is the yellowish crystalline substance sometimes found encrusted on eyelids when you wake up. There seems to be no widely used specific term, perhaps because the effect is seen as trivial and erratic. None the less, it is important. During the day grit, dead cells and other debris accumulate in the tears which are more than mere saltwater.

Mucoproteins cover the eyeball, curdling protectively around sharp grit to encase it in mucus, a middle salty layer is the main liquid part, and an outer, oily layer reduces evaporation. At night, movements of the eye and closed eyelids stir this orbital midden, massaging solids towards the inner corner of the eyelids. There the exposed liquid evaporates until the residual sludge forms pellets that you remove harmlessly by washing, or with your finger, the next morning.

Gritty environments such as deserts may damage eye tissues enough to convert your tears into dilute pus. This dries on the edges of your eyelids, gluing them shut in spite of the waxy coating that normally reduces spillage and keeps their epidermis water-repellent. It can be very disconcerting to awaken from an exhausted sleep to find your eyelids sealed shut so that you think it is still dark. If this ever happens to you, soak them open gently, or you may lose some eyelashes in the sand.

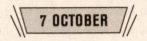

Why do submarines travel faster in colder water?

DENSITY AND VISCOSITY do not vary significantly over the range of temperatures typical of seawater, so differences in drag on the submarine cannot account for this. The efficiency of the vessel's propeller has more to do with it.

A spinning propeller creates regions of high and low pressure. Where the pressure falls below the saturated vapour pressure for a dissolved gas, the gas comes out of solution and forms bubbles. When the bubbles collapse, they create noisy shock waves; this is akin to what happens in a kettle just before it boils. These bubbles interfere with the action of the propeller on the water.

In warm water this bubble formation, called cavitation, happens at lower propeller speeds than in cold water. This is because the gas is more soluble in cold water and its saturated vapour pressure is lower. So a submarine can run silently at higher speeds in colder water.

The submarine's engine should also be more efficient in colder water, though this may not be very significant in practice. All engines that convert heat energy into mechanical work exploit the temperature difference between hot and cold reservoirs. The efficiency is given by the temperature difference divided by the temperature of the hot reservoir. Other things being equal, the temperature difference and hence the efficiency will be greater when the seawater (the cold reservoir) is at a lower temperature.

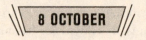

Why does garlic turn blue in vinegar?

Mix GARLIC CLOVES and apple cider vinegar, pop it in the fridge for a day or so and you'll notice that the garlic will turn bright blue. The discolouration is the result of some complicated chemistry involving the garlic's flavour compounds. The phenomenon is confusingly called 'greening', and the food industry has encountered enough accidentally coloured batches of processed garlic for it to have generated some interest.

In the traditional Chinese pickle of garlic cloves in vinegar known as Laba garlic, the colouration is intentional. Chemists have speculated on its cause since at least the 1940s, and in the last few years Chinese and Japanese researchers have worked out what is going on.

The flavour of garlic is generated when an enzyme called alliinase acts on stable, odourless precursors. These are normally in separate compartments in the cell but can combine if there is damage, including that caused by vinegar. The major flavour precursor in garlic is alliin (S-2-propenyl cysteine sulphoxide) while a minor one is isoalliin ((E)-S-1-propenyl cysteine sulphoxide).

Key to the colour change is a product of these reactions called di-1-propenyl thiosulphinate. It can react at slightly acid pH with amino acids from the ruptured cells to form pyrrole compounds, which are then linked together by di-2-propenyl thiosulphinates to form dipyrroles. These are reddish purple, but as the cross-linking continues, molecules with deeper and bluer hues are formed. Among these are compounds called phycocyanins, which are related to chlorophylls and are found in some algae that are used as blue colouring by the food industry.

Keeping garlic somewhere cool increases the amount of isoalliin present, which is why the best Laba garlic is produced several months after harvest. Isoalliin is also the major flavour precursor in onions. They smell different from garlic because they lack alliin and have a second enzyme that intercepts the product of the alliinase reaction to form onions' characteristic tear-producing molecules. Onions do not turn blue because this second reaction leaves less thiosulphinate to be converted to coloured compounds. This explains why onions undergo 'pinking' instead.

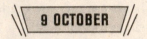

9 OCTOBER

Are most organisms bilaterally symmetrical?

WE TEND TO assume bilateral symmetry – having a body with two sides that are mirror images of one another – is normal because that is what we and most of the organisms we notice (vertebrates and arthropods) display. But bilateral symmetry is the exception rather than the rule; many creatures exhibit radial or even spherical symmetry. Some alter their symmetry over time – for example, a starfish will start out as a bilaterally symmetrical larva and become radially symmetrical as it matures.

Asymmetry is commonest among organisms that have little need of well-defined structures in their bodies. Some algae, fungi and sponges never developed much symmetry, while parasites can abandon symmetry when they grow opportunistically to secure food. An example of the latter is *Sacculina*, a barnacle that injects its soft body through a crab's shell and then grows a lump of reproductive tissue plus a tangle of feeding filaments throughout the crab's body. This creature has no need for symmetry.

Humans are not quite bilaterally symmetrical either: our liver is on the right side, our spleen on the left, while our right lung has three lobes and the left two. We even slip into fractal symmetry when it suits the purpose: take a close look at the capillaries which transport blood to the tissues. We are not even superficially symmetrical. Next time you get out of the bath ask yourself, 'Do they both hang the same?' This works for either sex.

We are all changed and shaped by both our genes and our environment. To put it another way, we all conform to a pattern while being eccentric. Heck, that's life.

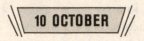

Is it possible get multiple colds at once?

THERE ARE MORE than 200 different viruses capable of causing the common cold in humans. And while there is little research into whether you can be infected with more than one common cold virus at the same time, there's no reason it shouldn't be possible. Our bodies are often invaded by more than one virus, which is why we can have a cold and a cold sore.

It's unlikely that the same cell could be infected by more than one cold virus, as cells produce a substance called interferon when under attack, which protects them from further invasion until that infection passes.

However, Rhinoviruses, responsible for up to 80 per cent of cases, infect only a few cells, and the infection usually involves only a small portion of the cells of the respiratory epithelium, leaving many cells available for infection by other viruses. So it seems it would be possible to catch several colds at once, each one being caused by a different virus targeting different cells. You would probably not be aware of such simultaneous infections, though you may feel worse than if you had only one infection, as more of the respiratory apparatus would be infected.

The body has basically two types of immune reactions to a viral infection: non-specific and specific. Once infected, your body will first use the non-specific system, which fights off the intruder using a general mechanism that is deployed against all invaders. If this is not effective and the virus persists, the specific system is used.

The specific system recognises the virus and produces specific antibodies. Recognition is usually based on identifying the complex molecules of the virus (proteins, glycoproteins or complex polysaccharides), otherwise known as antigens. Once the antigens are detected, the body teaches cells called B and T lymphocytes to fight them. Some of the new B lymphocytes turn into 'memory' cells which can survive for several decades, so the next time the antigen is detected, the system can respond immediately.

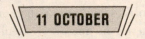

11 OCTOBER

Would more goals be scored in a game of football with two balls?

ONE CRITICISM OFTEN levelled at football is that the game can be unexciting as not many goals are scored in most matches. Would two footballs up the ante?

If the game were modelled as a simple two-dimensional system, with the football ricocheting off players and sidewalls, then the number of times a ball crosses the goal line would be a function of the number of balls. For example, the number of goals scored would double with two balls in play. In practice, however, the number of goals might well increase by a bigger factor. When in possession of both balls, an attacking team could use one ball as a decoy, increasing the chances of scoring with the other.

But, quite apart from the difficulty of refereeing such games, putting more than one ball into play would increase the risk of injury as players collide in an attempt to play different balls. It would be much safer to experiment with the offside rule. For example, the rule could be changed so that only attacking players entering the penalty area before the ball would be considered offside. This would have the added benefit of preventing defences playing the offside trap, which can diminish the game as a spectacle.

Or instead of having more than one ball, it might be entertaining to see more than two teams on the pitch. Tri-Soccer consists of one ball but three teams, each defending its own goal, but permitted to score in either of the other goals. Along these lines the pitch could be circular with goals arranged 120 degrees from each other.

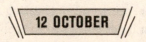

Why don't birds fall off their perch when they sleep?

BIRDS DO SLEEP, usually in a series of short 'power naps'. Not only that, but some do it standing on one leg. Swifts are famous for sleeping on the wing. Since most birds rely on vision, bedtime is usually at night, apart from nocturnal species, of course. The sleeping habits of waders, however, are ruled by the tides rather than the sun. The reason birds don't fall off their perch when they sleep is because they have a nifty tendon arrangement in their legs. The flexor tendon from the muscle in the thigh reaches down over the knee, continues down the leg, round the ankle and then under the toes. This arrangement means that, at rest, the bird's body weight causes the bird to bend its knee and pull the tendon tight, so closing the claws.

Apparently this mechanism is so effective that dead birds have been found grasping their perches long after they have died.

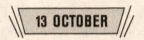

13 OCTOBER

Who is the voice in your head?

FOR SOCRATES, IT came as a warning when he was about to make a mistake. For Sigmund Freud, it was a loved one accompanying him when he travelled alone. Hearing voices has a long history.

And as those distinguished gents perhaps attest, it isn't always a sign of madness: our everyday thoughts often sound pretty voice-like. In 2011, Charles Fernyhough and Simon McCarthy-Jones of Durham University, UK, found that 60 per cent of us experience 'inner speech' with a back-and-forth conversational quality.

So where does inner speech end and hearing 'outside' voices begin? One answer is that an inner voice sort of feels like you, so you feel more control over it – but given how involuntary many thought processes seem to be, that is rather unsatisfying.

Voices aren't the only expression of our inner thoughts – our minds tell us stories, too. This 'confabulation' is a symptom of some memory disorders, whereby people have false recollections. But the rest of us do it too. Experiments show, for instance, that when people are forced to make a random decision they later invent a narrative to explain it.

One theory is that this helps us make sense of a world that bombards us with information, and gives conscious rationale to decisions we make for unconscious reasons. Robert Trivers, an evolutionary biologist at Rutgers University in New Jersey, thinks our lies are more self-serving: by lying to ourselves, we lie better to others too.

This may explain the phenomenon known as positivity bias,

whereby people overestimate their virtues. 'We put ourselves in the top half of positive distributions,' says Trivers. 'Eighty per cent of US high school students believe they are in the top half for leadership ability.' With these boosting voices, you probably shouldn't be too worried about what you hear – just don't believe everything they tell you, either.

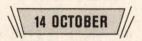

Why does lemon juice stop cut fruit from browning?

To ANSWER THIS question first we need to understand why some plant tissues go brown when cut. Plant cells have various compartments, including vacuoles and plastids, which are separated from each other by membranes. The vacuoles contain phenolic compounds which are sometimes coloured but usually colourless, while other compartments of the cell house enzymes called phenol oxidases.

In a healthy plant cell, membranes separate the phenolics and the oxidases. However, when the cell is damaged – by cutting into an apple, for example – phenolics can leak from the vacuoles through the punctured membrane and come into contact with the oxidases. In the presence of oxygen from the surrounding air these enzymes oxidise the phenolics to give products which may help protect the plant, favouring wound healing, but also turning the plant material brown.

The browning reaction can be blocked by one of two agents, both of which are present in lemon juice. The first is vitamin C, a biological antioxidant that is oxidised to colourless products instead of the apple's phenolics. The second agents are organic acids, especially citric acid, which make the pH lower than the oxidases' optimum level and thus slow the browning.

Lemon juice has more than 50 times the vitamin C content of apples and pears. And lemon juice, with a pH of less than 2, is much more acidic than apple juice as a quick taste will tell you. So lemon juice will immediately prevent browning.

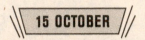

15 OCTOBER

Why does eating spinach make your teeth feel weird?

You CAN TRY this experiment at home; better yet, you might be able to convince the kids to eat their greens in the name of science . . .

Boil some spinach (canned works best, but fresh is fine) until it is cooked, drain it, let it cool a little and eat it. Then run your tongue around your teeth and mouth. Your teeth and the inside of your mouth will feel fuzzy and furry. This may be one of the reasons why children particularly dislike spinach, but once they realise they are part of a noble experiment their attitude changes radically.

Spinach contains a large amount of oxalate crystals – mineral salts of oxalic acid. When spinach is cooked – especially the canned variety – some of the spinach cell wall structure is damaged and the oxalate crystals leak out. These coat your mouth to give the fuzzy feeling, and it explains why fresh, uncooked spinach does not produce a similar effect. Spinach is also rich in calcium and oxalic acid, and these combine with the calcium in saliva to deposit large amounts of furry, calcium-rich plaque on your teeth. Chard and beetroot leaves have a similar effect.

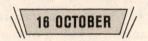

Do fish fart?

IN 2003, BIOLOGISTS linked a mysterious underwater farting sound to bubbles coming out of a herring fish's bottom. No fish had been known before to emit sound from its backside or to be capable of producing such a high-pitched noise. Apparently, it sounded just like a high-pitched raspberry.

Fish are known to call out to potential mates with low grunts and buzzes, produced by wobbling a balloon of air called the swim bladder located in their tummies. The swim bladder inflates and deflates to adjust the fish's buoyancy.

The biologists initially assumed that the high-pitched sound they had detected was also coming from the swim bladder in the fish's stomach, but then they noticed that a stream of bubbles expelled from the fish's bottom corresponded exactly with the timing of the noise . . .

Because the scientists needed a more important-sounding name than 'fish farts', they decided to call the noise Fast Repetitive Tick (FRT – which is a bit like fart). Unlike a human fart, the sounds were probably not caused by digestive gases because the number of sounds did not change when the fish were fed. The researchers also tested whether the fish were farting from fear, perhaps to sound an alarm, but when they exposed fish to a shark scent, there was again no change in the number of farts.

Three things persuaded the researchers that the fish farts were most likely to be produced for communication. Firstly, when more herring were in a tank, the researchers recorded more farts per fish. Secondly, the herring were only noisy after dark, indicating that the sounds might allow the fish to locate one another when they could not be seen. Thirdly, the biologists knew that herring could hear sounds of this frequency, while most other fish cannot. This would allow them to communicate by farting without alerting predators to their presence.

It's just a theory for now, but the discovery means that one day scientists might be able to track fish by their farts in the same way that whales and dolphins are monitored by their high-pitched squeals.

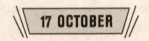

Could a racing car drive upside down?

IN THEORY, YES, an F1 car could drive upside down – but only in theory. Downforce acts towards the road, whatever the road's orientation, and it increases roughly with the square of the vehicle's speed. Driven fast enough, the downforce exceeds the weight of the car, which could then run along the ceiling of a tunnel. Depending on the set-up, the downforce and the weight of an F1 car typically become equal when the car is running at 130 kilometres per hour, though F1 cars are capable of generating up to three times their weight in downforce.

Filling a wardrobe with clothes boosts its weight. This increases the friction between it and the floor, making it harder to slide. An F1 car designer wants to increase the friction between the tyres of a car and the track so that it can carry more speed into corners without sliding off. But the designer wants to achieve this without increasing its weight.

So downforce is the answer and it can be achieved in two ways. First, upside-down wings are angled to deflect air upwards, away from the track, resulting in a reactive force on the car in the opposite direction. Second, designers exploit the Bernoulli effect. Pass air through a narrowing gap and it speeds up. This is what happens beneath an F1 car because the space between the ground and chassis represents a constriction to the airflow. According to Bernoulli's principle, this leads to reduced pressure under the car. The ambient pressure above the car is higher than that beneath it, leading to a net force in the direction of the road.

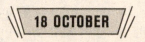

18 OCTOBER

Why can't you tickle yourself?

IF SOMEONE WAS tickling you and you managed to remain relaxed, it would not affect you at all. Of course, it would be difficult to stay relaxed, because tickling causes tension for most of us, such as feelings of unease due to physical contact, the lack of control and the fear of whether it will tickle or hurt. However, some people are not ticklish – those who for some reason do not get tense.

When you try to tickle yourself you are in complete control of the situation. There is no need to get tense and therefore, no reaction. You will notice the same effect if you close your eyes, breathe calmly and manage to relax the next time someone tickles you.

The laughter is the result of the mild state of panic you are in. This may be inconsistent with 'survival of the fittest' theories, because panic makes you more vulnerable. But as in many cases, nature is not necessarily logical.

Why don't all animals have round pupils?

IN ALMOST ALL vertebrates, the pupil is large and round in low light but constricts as light levels rise. Although the constricted pupil is round in most species, in some animals it can be a vertical or horizontal slit, or even, as in some geckos, composed of several tiny pinholes. The cuttlefish has perhaps the most exotic pupil shape: it becomes a W in bright light.

In low light the primary requirement is to be able to see anything at all. This means the eye must gather as many photons as possible, which requires a large pupil. However, there is a disadvantage to this, namely reduced image quality. This is because when the pupil is at its widest, the entire lens – lying behind the pupil – is used to form the image. Light going through the edges of the lens ends up being focused more than light going through the centre, an effect known as spherical aberration. Not all of the rays are focused at the same point, resulting in a blurry image. In low light however, this is a price worth paying. At high light levels the pupil constricts to ensure only a small part of the lens is used, resulting in better image quality.

In humans and other animals with round pupils, a sphincter muscle at the tip of the iris constricts to form the pupil. The disadvantage of a circular muscle is that the pupil cannot be closed down beyond a certain size. In animals with a slit pupil, such as a cat, the corresponding muscle runs along either side of the slit.

In comparison to humans, where the difference between the constricted pupil and dilated pupil differ in area by a factor of about 16, a cat's pupil changes by a factor of at least 135, and the pupil area can approach zero in bright light. Slit pupils are widespread in nocturnal animals whose retinas are adapted for low light and might be damaged in bright conditions.

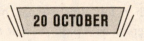

Why does time move forwards?

THERE IS A reason we say time goes by: it seems to flow. No matter how still we stand in space, we move inexorably through time, dragged as if in a current. As we do, events steadily pass from the future, via the present, to the past. Isaac Newton saw this as a fundamental truth. 'All motions may be accelerated and retarded, but the flowing of absolute time is not liable to any change,' he wrote.

So how does time flow, and why always in the same direction? Many physicists will tell you that's a silly question. For time to flow, it must do so at some speed. But speed is measured as a change over time. Even if time were standing still, it could be said that for every second that passes, one second passes. Indeed, if that's a measure of flow, we could say that space flows: it passes at one metre per metre.

Special relativity has revealed that there's no such thing as objective simultaneity. Although you might have seen three things happen in a particular order – A, then B, then C – someone moving at a different velocity could have seen it a different way – C, then B, then A. In other words, without simultaneity there is no way of specifying what things happened 'now'. And if not 'now', what is moving through time?

Rescuing an objective 'now' is a daunting task. But Lee Smolin of the Perimeter Institute for Theoretical Physics in Waterloo, Canada, has given it a go by tweaking relativity. He argues that we can rewrite physics in a way that includes 'now' if we sacrifice some of our objective notions of space.

Most physicists aren't having it. The general consensus is that time is more or less just like space – an immutable dimension, stretched out through a four-dimensional 'block universe'. 'Every moment in that universe has a past, present and future,' says Sean Carroll from the California Institute of Technology in Pasadena. 'A person is described as a history of moments, and those moments all have a feeling that they're moving from the past to the future.'

That doesn't answer the question so much as shift it. If time does not flow, what makes us think it does?

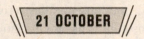

21 OCTOBER

Why does rain fog up the inside of a bus window?

IF YOU'VE BOARDED a bus while it's raining outside, you'll notice that the inside of the windows are often steamy. This is because the windows of the bus cool through contact with the cold air outside. Heat conducted through the glass (assuming it is single-glazed) will keep the inside face quite close to the outside temperature.

The inside face of the glass will steam up when its temperature is lower than the 'dew point' of the inside air – that is the temperature to which this air needs to be cooled to become saturated, so that moisture begins to condense out of it. In the case described, the dew point of the air in the train could be just above 7 °C.

On a rainy day the passengers' clothing will be wet and their umbrellas will be dripping onto the floor and dampening the upholstery. The heat inside the bus will cause the rainwater to evaporate, markedly raising the moisture content of the air. The dew point inside will therefore be higher. Consequently, more moisture will condense on the window on a rainy day than on a dry one. This effect will be exaggerated if windows are shut, trapping the moist air inside the bus.

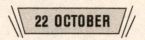

22 OCTOBER

Why does molten cheese go stringy?

THE BEST PART of eating a cheese toastie (or grilled cheese, depending on where you're from) is undoubtedly the stringiness of the cheese as you pull away post-bite. But what is the science behind this delicious quality of cheese?

Uncooked cheese contains long-chain protein molecules more or less curled up in a fatty, watery mess. When you heat cheese, the fats and proteins melt and if you fiddle with the fluid, the chains can get dragged into strings. Grab a bit of the molten cheese and pull, and you get a filament, in much the same way that you can draw and twist cotton wool into yarn.

You can do similar things with polythene from plastic bags by heating or stretching the plastic to curl or stretch the long-chain molecules. When the molecules are curled up, the plastic is softish and waxy. When they are stretched into fibres, the result is elastic and strong in the direction of the stretch, although it splits easily between the chains lying along the fibre.

Can you fall asleep with your eyes open?

SOME PEOPLE CAN sleep through anything. Earthquakes, gunshots, bright lights – nothing rouses them. But these are people who are already asleep. In 1960, Ian Oswald of the University of Edinburgh, UK, wondered how much stimulus someone could be exposed to while awake and still drop off. Would it even be possible to fall asleep with your eyes open?

Oswald first asked his volunteers to lie down on a couch. Then he taped their eyes open. Directly in front of them, about 50 centimetres away, he placed a bank of flashing lights. No matter how much they rolled their eyes, they could not avoid looking at the lights. Electrodes attached to their legs delivered a series of painful shocks. As a finishing touch, Oswald blasted 'very loud' music into their ears.

Three young men volunteered to be Oswald's guinea pigs. In his write-up, Oswald praised them for their fortitude. Yet, all the lights and noise and pain made no difference. Once they were tired, an electroencephalograph showed all three men to be asleep within 12 minutes. Oswald worded his findings cautiously: 'There was a considerable fall of cerebral vigilance, and a large decline in the presumptive ascending facilitation from the brain-stem reticular formation to the cerebral cortex.' The men themselves were more straightforward. They said it felt like they had dozed off.

Oswald speculated that the key lay in the monotonous nature of the stimuli. Faced with such monotony, he suggested, the brain goes into a kind of trance. That may explain why it's easy to doze off, even in the middle of the day, while you are driving along an empty road. How much this will help when sleep eludes you while you're stuck on a red-eye flight is another question. Asking the baby in the row behind you to scream more rhythmically is unlikely to do the trick.

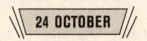

Why is the scrotum wrinkly?

THE SCROTUM PLAYS a valuable role in thermoregulation of the testicles, as sperm production is reduced at core body temperatures and above. People with this piece of anatomy may notice that on a very cold day, or when emerging from a cold shower, the scrotum (and the rest) will be much smaller. The converse happens on a warm day or on emerging from a hot bath. The scrotum is quite smooth when relaxed and only wrinkles as it is pulled tight to the body.

This is because muscles within the scrotal wall contract to bring the testicles nearer the abdomen in cold conditions, and relax to keep them cool when it is hot. There is also a complex system of vasculature in the spermatic cord (the pampiniform plexus), where warm blood leaving the abdomen is cooled by a heat exchanger system before entering the testicles.

When farmers first started to rear merino sheep in Australia, they found that the rams were sterile until sheared, so their testicles are shorn on a regular basis today. This breed of sheep originated in southern Portugal, where summer temperatures are not as high. Consequently, the rams have unsuitably woolly scrotums for southern Australian conditions.

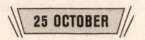

Why are cartons so tricky to pour?

MANY LIQUIDS POURED from cartons, such as orange juice, milk or soup, result in a sticky floor. And it's difficult to avoid the effect because when the carton is full you have little choice but to tip it gently when trying to fill your glass.

Try this for yourself with a carton of milk (or any other liquid) and a glass. At low pouring speeds the milk will cling to the edge of the carton and dribble its way down the container before depositing liquid on the floor – this is where a cloth may come in handy. At faster speeds the liquid will pour freely, allowing you to fill your glass with aplomb.

When the carton of liquid is tipped, the surface of the liquid in the container is raised and moves towards the opening of the container. As the carton is tipped further, liquid pours from the opening, creating pressure at the opening. In addition to this pressure force, there are surface tension forces acting on the fluid that tend to draw it towards the surfaces of the container. At high pouring speeds, the pressure force is much greater than the surface tension forces, and the fluid will leave the carton in an orderly fashion, following a predictable parabolic path towards your glass.

However, at low pouring speeds, a point is reached where the surface tension forces are sufficient to divert the path of the fluid jet so that it fails to leave the opening cleanly and becomes attached to the top face of the carton. Once attached, a jet of liquid will tend to stick to that surface thanks to the surface tension forces and a phenomenon known as the Coanda effect. This occurs when a fluid jet on a convex surface (such as a water jet from a tap curving around the back of a spoon) generates internal pressure forces that effectively suck the jet towards the surface.

The combined result of surface tension and the Coanda effect enable an errant flow of fluid to negotiate the bend from the top face of the carton onto the carton's side and, ultimately and rapidly, onto the floor – or your shoes.

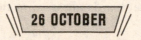

If you fired a gun into the air, how high would the bullet go?

IN MANY PARTS of the world, people celebrate victories, birthdays and similar events by firing guns into the air with great exuberance and a seeming disregard for the welfare of themselves and others. Assuming the barrel of the gun is perpendicular to the ground when the bullet leaves it, approximately what altitude would it reach and what is its velocity (and potential lethality) when it falls back to Earth?

As might be expected, measurements of this kind are rather difficult and the following values come from a computer model of the bullet flight. For a typical modern 7.62 millimetre calibre bullet fired vertically into the air from a rifle, the bullet will have a velocity of about 840 metres per second as it leaves the muzzle of the gun and will reach a height of about 2,400 metres in some 17 seconds. It will then take another 40 seconds or so to return to the ground, usually at a relatively low speed which approximates to the terminal velocity. This part of the bullet's trajectory will normally be flown base first since the bullet is actually more stable in rearward than in forward flight.

Even with a truly vertical launch, the bullet can move some distance sideways. It will spend about 8 seconds at between 2,300 and 2,400 metres and at a vertical velocity of less than 40 metres per second. In this time it is particularly susceptible to lateral movement by the wind. It will return to the ground at a speed of some 70 metres per second.

This sounds quite low but, because of the predominance of cranial injuries, the number of deaths and serious injury as a proportion of the number of gunshot wounds is surprisingly high. It is typically some five times more than is observed in normal firing. Don't try this one at home.

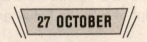

27 OCTOBER

Why does the brain have so many folds?

THE BRAIN HAS fissures to increase the surface area for the cortex. Much of the work carried out in the brain is performed by the top few layers of cells – a lot of the brain's volume is, in effect, point-to-point wiring. So, if you need to do lots of processing, it is much more efficient to grow fissures than it is to expand the surface area of the brain by increasing the skull diameter.

Another reason our brains need such a large surface area is down to heat. Brain tissues consume massive amounts of energy and the resulting heat that is generated has to be dumped. Put your hand on your head and feel how hot it feels compared to your thigh. Brains of lower vertebrate animals lack extensive folds because they have relatively less heat to get rid of.

Humans, on the other hand, have large brains which do a lot of work. The extra folds in our brains increase the surface area for blood vessels to dump the excess heat produced by all that hard thinking. If our brains were to evolve into more complex and larger organs, their folding would have to increase exponentially in order to be able to release the additional heat that they would produce.

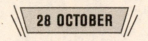

28 OCTOBER

Which is smoother, Earth or a squash ball?

To ANSWER THIS intriguing question, we first need to establish the scale factor we would have to shrink the Earth by, in order to reduce it to the size of a squash ball, so that an effective comparison could be made.

The Earth is 12,756 kilometres in diameter at the equator, and a regulation squash ball has a diameter of 4.4 centimetres. This means that to shrink the Earth down to the size of a squash ball, its size would have to be multiplied by a scale factor of 3.45×10^9.

To compare the smoothness of the two surfaces, we need to know the variation of the surface – that is, the difference between the highest and lowest points.

For a squash ball, this is a simple process, because there are very few areas where the surface is higher than the average, but there are many small indentations or depressions. Since these depressions are roughly 0.1 millimetres in depth, the variation in surface height can be taken to be roughly 0.1 millimetres, or 10^{-4} metres.

For the Earth, the lowest point below the surface is in the Mariana Trench, which is 11,034 metres below sea level at its deepest point, known as the Challenger Deep. The highest point is, of course, the summit of Mount Everest, which is estimated at 8,848 metres above sea level. Therefore, the variation in the height of the Earth's surface is 19,882 metres.

If we scaled the Earth down to the size of a squash ball, using the scale factor calculated above, the variation of its surface would be 6.86×10^{-5} metres, or 0.0686 millimetres. This figure is in fact about two-thirds of the figure for the squash ball, so if Earth were scaled down to this size it would indeed be smoother than the average regulation squash ball.

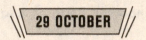

Why do plants with berries often have thorns?

IT MIGHT SEEM like such plants are sending mixed signals, but actually they have a very clear message: 'eat my fruit, not my leaves'. Most animals that benefit the plant by eating fruit and dispersing seeds are dexterous enough not to hurt themselves. A bigger, clumsier herbivore might not bother separating fruit from foliage, but if the latter came with a mouthful of thorns, the animal might think again.

A related issue is why some tasty fruits have highly distasteful skins – oranges, bananas and mangoes are good examples. To enjoy these, an animal has to be big and powerful enough to peel the fruit, and thus probably capable of eating the peeled fruit whole. This means swallowing the contained seeds and depositing them many miles away, rather than just nibbling the juicy bits and leaving the seeds unhelpfully close to the parent plant.

This is similar to the reason that unripe fruits are small, inconspicuously coloured, sour and rich in unpleasantly astringent tannins. The situation changes when the seeds are fully developed and could benefit from being dispersed.

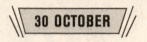

What is earwax for?

EARWAX, ALSO CALLED cerumen, acts as a cleaning agent for the ear with lubricating and antibacterial properties. Cleaning occurs because the epithelium – the surface layer of skin inside the ear – grows outwards, from eardrum to exit, acting as a conveyor belt carrying dust or dirt out of the ear.

At first, this migration is as slow as fingernail growth but, aided by jaw movement, it accelerates once it reaches the entrance of the ear canal. When it gets to the final third of the ear canal, where cerumen is produced, the conveyor carries both wax and whatever gunge it has accumulated towards the exit. Cerumen consists of a mixture of watery secretions from sweat glands and more viscous secretions from sebaceous glands. Some 60 per cent is keratin but it also contains dead skin cells, fatty acids, alcohol and cholesterol.

There is more to earwax than meets the eye. It helps anthropologists track patterns of human migration because your genes determine whether you have 'wet' or 'dry' earwax. The absence of water in dry earwax, making it grey and flaky rather than the more common moist golden variety, eliminates evaporative cooling, an advantage to those who evolved in cold climates. The gene that reduces sweat production in these people is also responsible for dry earwax.

Surprisingly, problems with earwax can kill. Scuba divers sometimes have difficulty equalising the pressure in the inner ear to the ambient pressure of the water, which increases as the diver descends. It is usually caused by blocked or narrow Eustachian tubes, but earwax can also be responsible. In such circumstances, one ear will usually 'clear' before the other. A difference in pressure in each ear induces vertigo, an alarming spinning sensation.

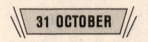

Do spiders drink water?

MANY SPIDERS, SUCH as the common garden spider, will devour their web first thing in the morning. In doing this, they consume the water that has condensed as dew droplets on the web. Other spiders, such as the whip spider, can use their pincers to take water into their mouths.

The black widow or the red back do not drink water at all. They get all the fluid they need from the juice sucked out of their prey. Tarantulas, on the other hand, like to drink water droplets that have collected on nearby leaves and foliage.

There are some creatures, including mammals, that do not drink. The name koala is derived from the Aboriginal word 'no drink'. Koalas get the fluids they need from eating the leaves of plants such as the smooth-barked eucalyptus.

NOVEMBER

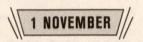

Why do pencil shavings smell so good?

AN IMPORTANT PROPERTY of a good-quality pencil is that it doesn't splinter when sharpened. Red cedar wood has this property and was the wood of choice for pencil manufacturers during the nineteenth and early twentieth centuries. Dwindling supplies of this tree forced manufacturers to turn to incense cedar in the 1940s and this is what many pencils are still made of today. Incense cedar has excellent 'sharpenability' and, as the name suggests, has a wonderful aroma.

Cedar trees have evolved to produce a cocktail of chemical compounds that includes cedrol and cedrene. This concoction can act as a protection against pests, bacteria and fungi. When a pencil is sharpened, this mix is released from the wood and the sun's heat increases the evaporation of the aromatics, enhancing the odour.

The limbic system of your brain is involved in aspects of olfaction, emotion and long-term memory. This helps to explain how, decades later, the evocative smell of pencil sharpenings can stimulate memories of your class and frame of mind when you first breathed in cedar's heady bouquet. A poor man's potpourri of pencil shavings and spices sitting on a radiator can elevate the spirit during winter's dark days.

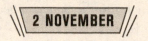

Do deserts exist at sea?

ARID AREAS OF ocean – where rain seldom, if ever, falls – certainly exist. They are caused by the circulation pattern of Earth's atmosphere.

In each hemisphere, between the latitudes of about 25 and 45°, there is a zone known as the subtropical high-pressure belt. This belt contains several separate high-pressure cells (also known as anticyclones), and moves north and south depending on the season, being about 5° closer to the equator in winter than summer.

This zone is the subsiding arm of the Hadley cell, a north–south circulation of air in the low latitudes, consisting of two opposing cells, each having air rising in the intertropical convergence zone (around the equator) and sinking in the subtropical high-pressure belt.

In general, the belt comprises vast areas of light winds and gently subsiding air. The subsiding stable air undergoes compressional warming, producing low relative humidities. The weather is usually fine and rain clouds few, resulting in arid climates over both land and sea. However, the far-western portions of the subtropical highs have less subsidence and the air is not as stable, so cloudy, stormy weather is more frequent there.

Most of the world's great deserts lie under the eastern flank of the subtropical anticyclones: the Sahara, the Kalahari, the deserts of the Southwest US and the Atacama in Chile, as well as vast areas of inland and western Australia. Large areas of ocean in the subtropical high-pressure belt are also arid.

These zones became known as the horse latitudes during the days of the great sailing ships. One explanation is that sailing ships becalmed in the belt would run short of water and, with no rain, horses being traded between Europe and the Americas would be thrown overboard, often into the Sargasso Sea.

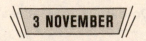

Why do birds have different songs?

BIRDSONG VARIES BY habitat type because the habitat has a profound effect on how these long-distance signals are transmitted. To minimise habitat-induced degradation, the acoustic adaptation hypothesis predicts that birds living in dense forests will have slower and more tonal calls, while those living in more open habitats will have faster-paced and buzzier calls.

Calls seem to be adapted to distance, noise, obstacles, habit and competition. The most elaborate singers inhabit open bush, where their song can convey complex information over long distances. In thick bush, only deep ventriloqual notes such as those of the ground hornbill carry for any distance. White-eyes foraging among dense leaves cheep softly, keeping flocks together at short range.

Even the apparently unsophisticated croaks, screams and yarps of seabirds vary in complexity and carrying power according to their habits and individual circumstances. When calling through wave noise over long distances they tend to screech shrilly, whereas when they are intimate they are quieter.

The effect is most pronounced when comparing contrasting habitat types, such as very open and very closed ones. Other factors, including the songs of species competing for acoustic space and the songs produced by closely related species, can also play a role.

Can you make plastic at home?

YOU WOULD IMAGINE that you'd need some pretty noxious, smelly chemicals to make plastic, but you can actually make malleable, doughy pieces of material in your own home. Instead of putting vinegar on your fish and chips and wasting your milk in your tea, use the two liquids to become a polymer chemist . . .

Pour a pint of milk into a saucepan and gently warm it. When the milk is simmering (don't let it boil) stir in 20ml of white vinegar until you notice whitish-yellow rubbery lumps beginning to curdle in the mixture at the same time as the liquid clears. Turn off the heat and let the pan cool.

As the vinegar is added and stirred, the liquid gets clearer and the yellowy rubbery lumps form. When the pan has cooled you can sieve the lumps from the liquid, tipping the liquid down the sink. Put on rubber gloves and wash the lumps in water. You can then press them together into one big blob – they will be squishy and will feel as if they are going to fall apart, but they will stick together after some firm kneading. You can now use your artistic skills to fashion the material into the shapes of your choice. Leave the material to dry for a day or two and it will be hard and plastic enough to paint and varnish.

You have used the combination of an acid – in this case vinegar, which contains acetic acid – and heat to precipitate casein (a protein) from the milk. Casein is not soluble in an acid environment and so, when the vinegar is added, it appears in the form of globular plastic-like lumps. Casein behaves like plastic because it has a similar molecular form. The plastics in everyday objects are based on long-chain molecules called polymers. These are of high molecular weight and get their strength from the way their billions of interwoven criss-crossing molecules tangle together.

The Indian cheese paneer is made in a very similar way to the plastic you have just made, although in this case lemon juice is the acid used rather than vinegar. Afterwards, unlike our plastic milk, it is not dried out and allowed to harden to tooth-breaking consistency, and so remains soft and edible.

How do sparklers sparkle?

THE SPARKS FROM the sparkler are produced by burning flecks of a metal such as magnesium or aluminium flung off from the firework. Initially only their outer layer burns, but after the fleck has burnt down to a critical size the core becomes so hot that it explodes. The sub-flecks from the explosion then burn out quickly and brightly in a distinctive star.

When you wave them around, they appear to draw lines. This is due to a phenomenon known as visual persistence. The human eye does not react instantly when its view changes, but keeps the old image around for a few milliseconds. This is what enables us to perceive films or television images as moving pictures when they are in fact a sequence of still images. The persistence of the eye causes each image to merge into its successor, creating the illusion of movement.

If the changing image contains very bright objects against a dark background – such as a sparkler at night – the persistence lasts longer, so the light from quite a long period of time can be added together to appear as a single streak.

There are numerous gadgets that exploit this effect by using strips of fast-moving LEDs to apparently create writing in the air. Persistence can also be seen in the coloured spots left in your vision after a camera's flash has gone off.

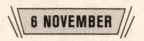

6 NOVEMBER

Why is there ash at the end of a bonfire?

LIKE MANY PLANTS, trees make sugars, cellulose and other organic molecules using carbon from the carbon dioxide in the air. They join it with hydrogen obtained by splitting water from their roots into hydrogen and oxygen. The oxygen is discarded into the atmosphere as part of this system of photosynthesis. All the other elements they need, like nitrogen, phosphorus and metals such as potassium, manganese, iron and zinc, are obtained with the water from the soil. When wood is burned, oxygen rejoins with the carbon and hydrogen from the organic compounds, releasing stored energy. Oxygen also joins with the trace elements, forming metal oxides and phosphates. It is these compounds that make up the solid ash, which is an excellent fertiliser, giving back nearly all the minerals originally taken from the soil.

Unfortunately, the nitrogen returns to the atmosphere. This explains why slashing and burning trees and other plants to create fields produces good yields for only a year or two before the land becomes nitrogen deficient.

343

Why does picking a chickenpox scab leave a scar?

HEALTHY CHICKENPOX SCABS can in fact leave inconspicuous scars, but the more you scratch them, the worse the scar will become.

Healing any clean wound starts with a growth of scaffolding tissue to contain the damage. Next the scar tissue begins adjusting its structure to its functions. In small, clean lesions of simple shape, the scar tissue may adapt so neatly that one hardly notices the mark that is left behind. However, scaffolding tissue cannot form so neatly on larger wounds, so they leave more conspicuous scars that may take years to shrink or may require plastic surgery.

Interfering with scar formation, by repeatedly scratching a scab, for instance, aggravates scarring. Also, when the tissues are infected with germs such as the chickenpox virus mentioned in the question, the pathogens not only interfere with tissue growth, they attract the leucocytes that can form pus. Left alone, healthy leucocytes cleanly kill both the pathogens and infected tissue; they cordon off the pustule until everything dries and sloughs off, so that scar tissue can form neatly afterwards. But interference with this process, such as scratching, messes up the pustule structure, exposes more tissue to pathogen and leucocyte damage, and thereby creates a larger and more unsightly scar.

How do multiple-headed flowers form?

THE TIP OF a plant shoot, known as the apical meristem, consists of undifferentiated cells with the potential to produce stems, leaves and flowers from a single growing point. When the pattern of growth is disrupted, the growing point can give rise to multiple fused stems. This is called fasciation (a word that comes from the Latin *fasces*: a bundle of sticks).

It is possible for entire plants to be affected, or for trees and shrubs to develop flat, paddle-shaped branches with dense clusters of leaves at the edges and tips. This disrupted growth can be caused by physical damage as a result of insect attack, fungi, bacteria and poor growing conditions, but the main cause seems to be genetic mutation.

A lot of media excitement was generated recently by a photograph showing double-headed Shasta daisies growing near the damaged Fukushima nuclear facility. Radiation might conceivably have caused this mutation, but such flowers are by no means uncommon.

The condition can be inherited, and in the UK it's quite common to see double-headed dandelions growing close together, probably the progeny of a single plant. Indeed, the variety of amaranth known as cockscomb (*Celosia*) has been selectively bred for its gaudy ruffled flower heads.

Another example of a plant with the mutation is the fantail willow, coveted by Japanese flower arrangers for its twisted, blade-like branches.

How can a helicopter achieve a loop-the-loop?

IN THEORY, MOST helicopters can perform a loop. With enough speed and distance from the ground they can use their momentum to do a loop and overcome the downward force produced by the rotors. During a properly flown loop centrifugal force is greater than gravity, even when the aircraft is inverted. The pilot is pushed into the seat throughout and experiences positive g-forces.

However, if the loop is too large or it is flown at too low a speed, the pilot falls away from the seat, restrained only by a harness, and experiences negative g-forces. In older helicopters the rotor blades are flexible and hinged, flapping up and down, but under positive g-forces they always flap upwards. Under negative g-forces the blades may bend downwards far enough to hit the tail of the helicopter, with fatal results. This is why loops are discouraged.

Modern military helicopters have stiffer, hingeless, rigid rotors, giving much greater agility. Even under negative g-forces the rotor blades remain a safe distance from the tail, allowing loops to be flown safely.

Remote-control helicopters can actually fly upside down. They do this by adjusting the angle of attack of their rotor blades (this is called collective pitch) so that they are in the opposite position to normal flight. They can get away with it because the forces acting on a small helicopter are lower and the joints and mechanisms are normally less complicated.

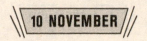

Does fizzy water weigh less than still water?

FIZZY OR CARBONATED water is heavier – that is, denser – than non-carbonated water if the carbon dioxide is in solution rather than forming bubbles.

For a given weight of water and dissolved carbon dioxide, the volume can be calculated by adding the volume of the water alone (about 1 millilitre per gram of water) to that of the gas (about 0.8 millilitres for each gram of carbon dioxide). So a solution of, say, 2 grams of carbon dioxide in 998 grams of water would have a mass of 1 kilogram, a volume of 999.6 millilitres and a density of 1.0004 grams per millilitre (at 4 °C, the temperature at which water is at its most dense).

However, if the water is fizzing, then it will be lighter (less dense) than still water. It is the bubbles that reduce the density: 2 grams of carbon dioxide gas would have a volume of about 1 litre.

Lake Nyos in Cameroon has carbon dioxide seeping into it from below. Usually the carbon dioxide-laden water stays at the bottom of the lake because it is denser, but sometimes it begins to form bubbles which cause it to rise. This sets in train a process that brings up a massive amount of carbon dioxide, which bubbles into the air. In 1986, 1,700 people were killed by carbon dioxide that escaped from the lake and smothered the surrounding valley.

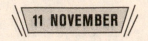

Why do we get chapped lips in winter?

COLD WINTER AIR cannot hold much moisture, so when it is warmed indoors to a comfortable temperature it will be drier than summer air of comparable warmth. This causes the skin to dry out more readily, rendering it less elastic and more prone to chapping.

Skin also expands and contracts with rising and falling temperature. This means the greater difference between indoor and outdoor temperatures in winter increases the stress placed on the skin as you go in and outside. Lips are particularly prone to chapping because they lack the oil glands present on the rest of the skin whose sebum acts as a natural barrier to desiccation and helps to keep it supple. Repeated wetting of the lips through licking only makes things worse by stepping up evaporative cooling.

Lips are also frequently flexed in the course of eating, talking and changing expressions, and, in an already dried state, acidic and salty foods only inflame them more. Smiling provides the final insult because it pulls the corners of the mouth considerably, stretching lips to breaking point.

If this all sounds familiar, then be reassured that you are far from alone in this, and the problem is relatively easy to remedy. Start applying lip balm or moisturiser regularly before the winter really sets in, and aim to wear a scarf that covers your mouth.

It is the same dry winter air that aggravates sinus headaches and raw coughs. Some remedies are simple – many householders find that humidifiers hooked over their radiators or having pot plants in the home provide relief.

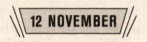

Why is the Large Hadron Collider so large?

THE LARGE HADRON Collider (LHC), the subterranean particle accelerator located just outside Geneva, Switzerland, is a synchrotron, a circular accelerator that uses carefully synchronised electromagnetic fields to accelerate particles to very high speeds. The faster the particles are moving, the more likely you are to see something interesting happen in a collision. So it's important to accelerate the particles, mainly protons, as much as possible. The protons need to follow a circular path so they can be continuously accelerated by an electric field, and this is done using magnets positioned around the tunnel. The faster the protons travel, the stronger the magnetic fields need to be to keep them on track.

To increase energy there are two possible choices: make the magnets stronger or the accelerator ring larger, so that the particles' path does not need to be bent so much. At some point there is either a technological or financial limit on the strength of the magnets, leaving ring size as the only remaining variable.

It is not just the LHC's circumference that is huge; so are the detectors. It's here that the collisions take place and physicists look for new particles. A detector is packed with all kinds of instruments to measure mass, energy, temperature and so on, and there are additional trackers and triggers right around the beam. In part these explain the detectors' size.

But there is something else. We know that a new particle can be recognised by the way in which it decays into particles we already know exist.

Take the Higgs boson, the long-sought particle that gives all other fundamental particles mass. It was discovered independently by two experiments at the LHC, known as ATLAS and CMS, in 2012. One of the ways the Higgs boson decays is into muons, which are massive relatives of the electron. We can see muons by tracing their paths. But they are very elusive particles and will fly through any substance with enormous speed, so the muon detectors at ATLAS, for instance, extend from a radius of 4.25 metres from the axis of the proton beam out to 11 metres.

If the detectors were closer to the beam or narrower than this, the muons would fly past before they could be identified. So the size of the LHC has much to do with what we can physically measure with the means available to catch the unseen wonders of nature.

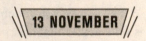

13 NOVEMBER

Which dinosaur would have evolved to be the most intelligent?

DINOSAURS DIED OUT some 65 million years ago and mammals exploited the vacant niche. If they had not died out, which branch of dinosaur might have evolved to become the dominant, most intelligent creature?

Birds evolved from theropod dinosaurs during the Jurassic period and species that survived the extinction event 65 million years ago are ancestors of those we see today. Were it not for the extinction event, the 'terrible lizards' might well still dominate. In the absence of environmental change, there is little selective pressure to drive the evolution of new species.

The climate change associated with the end of the Cretaceous period would have fired the starting pistol for an evolutionary race, with some species undergoing significant adaptive changes to fill the new niches.

This is akin to shuffling a deck of cards. It is impossible to work out in advance what species might dominate or even what a new species will look like, because it is impossible to know what combination of traits will be more favourable than another. Besides, sometimes it is enough just to be the first to occupy a particular niche.

14 NOVEMBER

Can a storm reverse a compass?

IN THE NOVEL *Moby-Dick*, the wooden whaling ship meets a typhoon south-east of Japan and is subjected to thunder, lightning and displays of St Elmo's fire. Subsequently, the magnetism of the ship's compass needle is discovered to be reversed. Author Herman Melville maintained that such compass reversals 'have in more than one case occurred to ships in violent storms', and sometimes when the rigging has been struck by lightning the magnetism in a compass needle may be totally lost. Is this fact or fiction, and if true how does it occur?

It turns out Herman Melville's assertion is entirely plausible. A moving electrical field will induce a magnetic field, and an electrical discharge such as lightning can easily cause a compass needle to lose or reverse its magnetisation.

Indeed, in the novel Captain Ahab fashioned a new compass by striking a sailmaker's needle and thereby magnetising it. This is rooted in fact. Ferromagnetic materials are composed of microscopic magnetic domains, which may be oriented in random directions producing a demagnetised state. By aligning the domains in more or less the same direction, the material becomes magnetised. In some cases, a sharp blow will impart enough energy to make this happen.

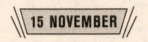

How long does it take to digest food?

HERE'S AN EXPERIMENT that kids will love – the gross factor looms large. All you'll need is your digestive system and some food. For the best results try sugar, streaky bacon, sweetcorn, tomatoes, mushrooms, celery, beetroot or any other fibre-rich vegetables.

Over the course of your next few bowel movements, take a closer look. Plant material such as red tomato is often visible in human excrement, as are the skins of peppers and capsicums. Sweetcorn is probably the most obvious retained product of the human digestive system, its yellow colour standing out particularly well in faeces, although celery and mushrooms are also difficult to break down, while beetroot can give a fascinating reddish hue, as it does to your urine.

The journey begins in our mouth where the food is crushed until the pieces are small enough to swallow, and enzymes begin to break it down further. Once in the stomach, it is attacked by more enzymes and acids in gastric juices and is blended by the churning motion of the muscles in the stomach wall. A couple of hours later this semi-liquid passes into the small intestine, from where much of it is absorbed into the bloodstream.

Some components, such as dietary fibre, cannot be digested and move on to the large intestine. This takes about six hours. Once there, more water is absorbed by the body so that only the indigestible material remains. This may take up to 36 hours before the remaining waste is expelled through the anus. The final faecal form is approximately 75 per cent water. The rest is bacteria, excreted biliary compounds and unfermented fibre.

Of the various foodstuffs we eat, sugar is the first to be absorbed, followed by proteins (from eggs, nuts and non-fatty meat) in about six hours, and finally various types of fat over longer periods, although the age, health and size of the eater will affect all these timings.

As you will discover from carrying out the experiment, some components of food, such as fibre, are hardly broken down at all and, as in the case of sweetcorn, pass out of the body relatively untouched. The lignins in mushrooms and the cellulose in celery are also relatively unscathed after a journey through your body.

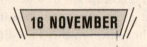

16 NOVEMBER

Why is microphone feedback high-pitched?

SOUNDS CAN BE viewed as combinations of sine waves of different frequencies. The Barkhausen stability criterion states that any frequencies that perfectly fit in the system's 'distance' (from microphone to speaker and back) and are amplified along the way will be sustained and amplified further.

This distance is hard to pin down because it depends on factors like delays in the electronics, the room's acoustic properties, the positions of the microphones and the resonant frequencies of the instruments and speakers. But higher frequencies are more likely to enter a feedback loop because the waves are shorter. The odds are higher that you can perfectly fit many short waves into a certain distance than many long waves. So although you do get throbbing low-frequency feedback, high-frequency feedback is more likely.

Moving a microphone near a speaker causes feedback because live speakers are never truly silent: close up, you can hear a multi-frequency hum. And as you move the microphone, you're changing the distance and sweeping the range of frequencies that cause feedback, resulting in a high-pitched whine.

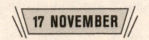
How far beyond the visible spectrum does a rainbow extend?

THE VISIBLE SPECTRUM runs from about 400 nanometres (violet) to 700 nanometres (far red). The spectrum produced by a rainbow will depend on two factors: absorption by water, mainly in the form of vapour (because light must pass through water droplets to form a rainbow) and the light source, in this case direct sunlight.

Water transmits light best at 400 nanometres. That is why everything looks blue underwater – the longer wavelengths are being absorbed. Water transmits reasonably well in the ultraviolet, up to 200 nanometres, so the strength of the violet end of a rainbow depends on the light source. As it happens, ultraviolet light from the sun is absorbed by the ozone layer and then dispersed in the atmosphere by what's known as Rayleigh scattering. So direct sunlight reaching Earth is relatively low in ultraviolet, with essentially nothing coming through below 300 nanometres. This then marks the short end of the rainbow's spectrum.

The infrared end of the rainbow is more to do with fading out. As light's wavelength increases from 700 to 1,000 nanometres, the proportion transmitted by water drops by 90 per cent. The available light from the sun is also down to half of its peak at 1,000 nanometres, so we can take that as a practical upper limit. There will be a couple of dips in the spectrum before that, appearing as dim bands in the rainbow. These are the result of absorption by atmospheric oxygen (at 762 nanometres) and water vapour (approximately 900 nanometres). In practice this gives us a rainbow spectrum ranging from 300 to 1,000 nanometres, although the infrared end will be quite faint.

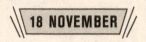

Why does sound fade over a distance but light does not?

SOUND WAVES FADE over distance. However, we can see light waves arriving from billions of light years away. Why don't incoming light waves dissipate before they reach us?

The answer stems from the fact that light energy is not dissipated as it travels through the vacuum of space, whereas sound cannot travel through a vacuum and needs a material in order to propagate.

When sound or light waves pass through material, the matter takes a toll. Even in transparent material, light is absorbed and the amount lost often depends on the light's wavelength. In the ocean, for example, a large fraction of the red end of the visible spectrum dissipates within about 10 metres of the surface – and not much light of any wavelength reaches depths greater than 50 metres even in very clear water. By contrast, sound usually suffers less attenuation through water than light does. Certainly a submerged scuba diver will hear the sound of an engine and rotating propeller even if the boat responsible cannot be seen.

Unless it is absorbed by an object blocking the way, sound or light from a point source spreads out in all directions and its energy expands over the surface of an imaginary sphere which gets bigger with time. So part of the reason why sound (or light) seems to fade with distance from the source is because its energy is being spread over a greater surface area. Of course, if stars and galaxies did not radiate such prodigious amounts of light energy they would not be visible from billions of light years away.

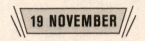

Can you get a cold from being cold?

THERE IS MUCH folklore about catching a cold when you're cold – but is there any connection between the two? And if not, why do many people still believe there is?

Science shows that there is, in fact, no connection. There is actually less chance of your catching a cold in the cold, because cold viruses die in cold temperatures.

The viruses that cause colds spread faster in the winter because people spend more time inside, where they are closer together. People close the windows in winter so air contaminated by virus particles is not diluted by 'fresh' air from the outdoors. This makes it easier for the virus to spread. Moreover, the cold, dry air of winter makes the mucous membranes in the nose swell. This produces the 'runny nose' we often incorrectly associate with an infection caused by a cold virus.

The origin of the old wives' tale that predicts colds, flu or pneumonia after being exposed to cold temperatures is the short period of fever that precedes the distinctive symptoms of these illnesses. These periods of fever make the patient feel cold and shivery. Shortly after developing other symptoms, the patient then associates the illness with having 'caught cold'. Indeed, the flu is called influenza from the belief that it was caused by the 'influence' of the elements. The fact that isolated researchers living in Antarctica never catch colds confirms that these are caught from people and not from 'cold'.

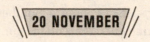

20 NOVEMBER

How many times do people fart a day?

HERE'S A PIECE of trivia with which to impress your friends: the average adult in the Western world farts roughly ten times a day, releasing enough gas to inflate a party balloon.

More than 99 per cent of these emissions are made up of five odourless gases. What exactly causes their foul smell has long been a matter of debate. But in 2001, one man believed he had the answer.

Michael Levitt, a gastroenterologist at the Veterans Administration Medical Center in Minneapolis, had been studying farts for over thirty years and solved numerous mysteries. As a result, he has become famous and his work has been written about in newspapers around the world.

Many scientists would welcome this exposure, but for Levitt it was disastrous. Readers would write angry letters to his employers complaining that his research was a waste of money. The temptation to turn out stories brimming with puns and fart jokes was also often too great for journalists to resist. One even called him Dr Fart.

But Levitt's work was serious. For instance, he was the first to correctly identify the gases that make farts smell. Evaluating a smell is a difficult task, so Levitt turned to the noses of two people with a rather unusual ability. Both could identify different sulphur-containing gases purely by smell. These lucky individuals were asked to evaluate the farts of 16 healthy men and women who, the previous evening, had eaten 200 grams of beans to ensure ample gas production. Levitt said the results pointed to hydrogen sulphide as the culprit in smelly farts, accompanied to a lesser extent by other sulphur-based gases.

Levitt's work was far from pointless – it has saved lives. Hydrogen and methane, two of the main gases that form in the gut, are combustible. In the 1980s, they caused a number of fatal explosions during otherwise routine operations on the gut. Somehow the purgatives used to clean the gut enhanced the production of hydrogen or methane and a chance spark during the operation triggered an explosion. Levitt and others have since developed purgatives that clean the bowel with minimal gas production. Thanks to Dr Fart, gut explosions are now rare.

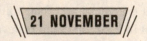

21 NOVEMBER

Would an object at the North Pole weigh the same at the equator?

How MUCH AN object weighs depends on the gravitational acceleration it experiences, so one enterprising *New Scientist* reader wanted to know the answer to this hypothetical scenario: if you were to buy 1,000 tonnes of gold bullion in Antarctica and then sell it in Mexico at the same price per kilogram, would you make a loss because the Earth is spinning faster?

Gravitational acceleration does indeed vary slightly over Earth's surface, partly through a combination of the planet being oblate (not perfectly spherical) and variations in the forces alluded to in the question. There are also more localised contrasts caused by differences in rock density and large-scale topographical features – the sea level around Greenland for example is higher than it would otherwise be, because of the large mass of water contained in glaciers. Further minor short-term fluctuations in apparent gravitational acceleration occur because of the relative motion of the sun and moon.

The Earth's spin reduces this acceleration by about 0.3 per cent at the equator compared with the poles. On top of this, the bulge of the planet at the equator means that objects here are further from Earth's centre than they are at the poles, giving another reduction of 0.2 per cent. An object will therefore weigh about 0.5 per cent more at the poles. Mexico City is at an elevation of more than 2,000 metres, which pushes it still further from Earth's centre. However, any additional effect is cancelled out by the fact that Mexico is still some distance from the equator. We can thus estimate that 1,000 tonnes of gold at the poles would reduce to a weight of 995 tonnes at the equator.

So in theory, a sizeable profit could be made if you were to buy at the equator and sell at the poles. However, in practice, the cost of transportation and security would be prohibitive. And the laws of supply and demand suggest that you wouldn't get a good price when penguins or polar bears are the only customers. In addition, gold bullion is usually cast in ingots of a certified mass – so unless you could persuade your customer to buy on the basis of the values on your own scales you would be in trouble.

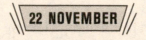

22 NOVEMBER

Why does your own snoring not usually wake you?

THERE ARE SEVERAL reasons why you may be the last to be bothered by your snoring.

First, one snores most loudly when deeply asleep and hardest to arouse. We live in bodies so noisy they interfere with our reception of external information, and we are equipped to ignore our own noises such as breathing. We subconsciously subtract such noises from the signals we hear in order to deal with our world.

Our signal-filtering processes can have peculiar side effects, as anyone can tell when hearing a recording of their own voice. Not only is the sound unfamiliar, but even the accent. Sound cancellation enables us to sleep through our own snores, whereas a bedfellow's snore or the merest rustle might arouse us. However, even our own snoring awakens us if a recording is played back or an inadvertent grunt breaks its rhythm in a way our 'cancellation software' cannot neutralise.

Even without waking us, severe snoring commonly interferes with sleep quality because of noise and airway interference. Research shows that many snorers, not only adults, forfeit healthily deep sleep. Some decades ago, staff at the Red Cross War Memorial Children's Hospital in Cape Town, South Africa, showed that children suffering from snoring, or sleep apnoea, benefited from positive-pressure air supplies that held the respiratory passages open.

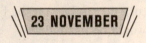

23 NOVEMBER

Why don't motorcycles fall over when taking corners?

WATCH A MOTORCYCLE race, and you'll observe that the motorcyclists taking corners at phenomenal speeds, often while titlting at well over 45 degrees from the vertical. Most of the time, they will manage this without sliding and crashing. How?

Turning a corner, a motorcycle is forced outwards by centrifugal force as well as downwards by the force of gravity. If the turn is taken with bike and rider too upright, centrifugal force flips the bike outwards and throws the rider off. If the bike leans too much, gravity makes it lie down and the tyres lose grip. It then slides out, with the rider usually sliding along behind.

With the bike leaning over at the best cornering angle, the combined forces push the mass of the bike out and down through the contact patches where the tyres touch the track.

Car tyres only need to work while upright. They have a square cross-section and there is tread only on the crown, not the sides. However, a motorcycle tyre has a rounded cross-section and the tread extends onto the sides of the tyre. This allows the tyres to continue to grip the track, even when the bike leans over in a corner.

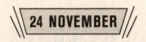

Is there a point on Earth where you can see both the Atlantic and Pacific oceans?

HERNANDO 'STOUT' CORTÉS is said to have climbed a tree on the Isthmus of Panama and seen both the Atlantic and the Pacific oceans from his vantage point. Are there actually any locations on the isthmus (or elsewhere) where this is possible?

The Isthmus of Panama is, at its narrowest point, 61 kilometres wide. From the middle of the isthmus, the horizon would have to be at least 30.5 kilometres away for both the Atlantic and Pacific oceans to be in view from a single vantage point.

The necessary height of the vantage point can be calculated from trigonometry, by mentally constructing a right-angled triangle, with a hypotenuse that is 30.5 kilometres long, atop a sphere of the Earth's diameter. A horizon 30.5 kilometres away can be seen from any vantage point which exceeds 85 metres above sea level.

The Times Atlas of the World tells us that near the centre of the Isthmus of Panama there are several geographical features which exceed 85 metres above sea level. The prime candidate is a ridge in the Cordillera mountains running roughly east to west and which is parallel to the Chagres river. There is a clear historical record that the earliest explorers of Panama could and did see the Atlantic and the Pacific from the same vantage point.

New Scientist readers have also informed us that there are various other places the Atlantic and Pacific oceans can be viewed at the same time: Costa Rica's highest peak, the 3,820-metre-high Mount Chirripó; Cape Horn at the southernmost tip of Chile; and the tip of the Trinity Peninsula on the continent of Antarctica should be fine for viewing both oceans too.

25 NOVEMBER

Which foods last the longest?

IN 1951, MRS E. Burt Phillips of West Hanover, Massachusetts, returned a 56-year-old can of clams to the manufacturer ('still edible', the press duly reported). A year later a 70-year-old crock of butter ('still white and sweet') was retrieved from an abandoned well in Illinois. In 1968, Sylvia Rapson of Cowley, UK, found a loaf of bread baked in 1896, still edible, tightly wrapped in table linen in an attic trunk, and when a house in Grimsby was razed to the ground in 1970, the ruins miraculously yielded up a 1928 packet of breakfast cereal.

But if you found a bit of food this old, would you eat it? In 1969, a man called George Lambert did just that. He turned up at the New Mexico state fair wearing his uniform from the 1898 Spanish–American war. Inside his mess kit he found a piece of hard tack, a long-lasting biscuit, and to the crowd's awe he bit a piece off and ate it. 'Tastes just like it did then,' the grizzled veteran announced. 'Wasn't any good then and it isn't now.'

Oscar Pike, a food scientist at Brigham Young University in Provo, Utah, has conducted taste and odour tests on everything from 30-year-old dried milk to oatmeal, Pike announced his results: the oatmeal wasn't all that good, but not all that bad either. When Pike was asked which old food he would eat himself, he said '30-year-old wheat. Baked into wholewheat bread it has practically the same sensory quality as bread made from freshly grown wheat.'

Grains are known for their longevity. In March 2006, 352,000 vitamin-fortified crackers from the Cold War turned up in a forgotten vault below the Brooklyn Bridge. A New York bridge inspector, sampling one, said they had 'a unique flavour'.

If you did eat a cracker as old as your grandfather, would you get sick? In the case of dried foods, the answer generally seems to be no. But then, why would you want to?

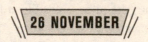

Can a sundial work on a cloudy day?

ONE WAY TO work out the position of the sun – and therefore to deduce the time – when conditions are overcast is to observe the polarisation of what light is available. This phenomenon, where the light waves oscillate in the same plane, is something that insects and birds exploit for navigation.

In general, scattered light is polarised at right angles to the sun. So when the sun is at its highest point, light is close to being horizontally polarised along the entire horizon. When the sun sets directly west, the sky will be vertically polarised at the horizon due north and south.

In 1848, English inventor Charles Wheatstone presented the 'polar clock', a sundial-like device that could be used when it was cloudy. By angling the tube towards the North Pole and turning a prism in the eyepiece until the light vanished, the relative angle of polarisation of available daylight could be deduced, giving the position of the sun and thus the approximate time. It has also been suggested that

Vikings used crystal sunstones to locate the sun when it was obscured by clouds or just over the horizon.

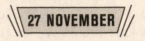

Why does orange juice taste so bad after you brush your teeth?

YOU KNOW HOW it is. You got up late and you're in a rush. You swig a glass of orange juice right after you've brushed your teeth. Instead of a sweet fruit flavour, the juice tastes like a foul bitter concoction.

You've just experienced the 'orange juice effect', something that's been known about since the 1970s. Foaming agents, usually the sodium lauryl sulphate, are added to toothpaste to help disperse the paste around your mouth, and for easy rinsing, but they have a nasty after-effect. They interfere with the taste buds on your tongue, suppressing the ability to taste sweetness and salt, and enhancing any bitter flavours. The powerful mint flavour that most of us want from our toothpaste simply adds to the problem. While it doesn't interfere with our taste buds in the same way as the foaming agents, the strong mint flavour overpowers the taste of anything that is consumed afterwards, including the breakfast you were looking forward to. The rule – and it makes sense for the health of your teeth too – is brush your teeth after meals, not before.

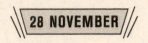

What is an exoplanet, and how many are there?

PLANETS THAT ORBIT around stars other than the Sun are known as exoplanets. They're tricky to see with telescopes because they are often hidden by the bright light of the stars they orbit.

Only a small percentage of exoplanets have been observed directly. The existence of others has to be inferred usually from observations of the host star. When a planet passes in front of its star as seen from Earth, the star appears to dim slightly. The measurement of these dips in brightness, through a technique called transit photometry, can yield a lot of information like the planet's orbital period and its size relative to its star. It also makes it possible to study its atmosphere and thus whether it might be home to extraterrestrials.

For an exoplanet to be observed in this way the planet, its star, and Earth all need to fall on the same plane. The probability of being able to detect such an alignment depends on the size of the star and the diameter of the planet's orbit. For a planet orbiting a sun-sized star at the same distance we are from the sun, the chance is around 1 in 200 that we would be able to detect it using transit photometry.

However, many other techniques can be used to hunt for extrasolar planets. One of the most successful is the radial velocity method. As a planet orbits, it will pull on its star, so the star is seen to move in a tight circle. Its light is blue-shifted while it is moving towards Earth, and red-shifted while it is moving away, and this can be detected using Doppler spectroscopy. At the time of writing, nearly 5,000 exoplanets have been found, well over 90 per cent of them by one of the two methods described above.

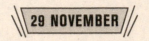

Why do you get a dry throat when you're nervous?

IF YOU HATE public speaking, you've probably experienced that awful moment when your throat clams up just as you're about to speak. This doesn't seem like a very useful response to nerves, so why do our bodies do it?

You get a dry mouth during public speaking because when you are nervous the body is set into the 'fight or flight' state. This is caused by an activation of the autonomic nervous system. It is seen throughout the animal kingdom, and has evolved to help the animal deal with dangerous situations – when escaping from predators, for example. You don't need to digest your last meal if a lion is trying to make you his next.

The nerves are selectively activated, depending on how important they are for the response. Because eating is not considered to be important at this time – you want to get the hell out of the place – the nerves to your mouth that control the salivary glands are suppressed, so your mouth dries up. In addition, your pupils dilate and the blood vessels to your muscles and heart are enlarged in order to get the blood to the most important organs needed for whatever drastic action is necessary.

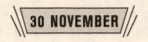

Can it be too cold to light a fire?

A FIRE IS a rapid exothermic (heat-producing) chemical reaction which occurs between a fuel and an oxidant, usually oxygen. The rate of any chemical reaction increases with temperature, as the molecules move faster and collide more often. Most fuels and oxidants can coexist at room temperature without spontaneously igniting. Although they may slowly react together, the rate of reaction is so slow that the heat produced is dissipated before the mixture can heat up – think of a slowly rusting iron nail.

To start a fire, you need to heat the mixture to its ignition point. This is the temperature at which the rate of reaction is high enough to produce heat more quickly than it is lost to the surroundings. Heat energy starts to accumulate, driving up the rate of the reaction, producing even more heat, and so on. This runaway reaction is what we call a fire.

So the answer to whether it can ever be too cold to start and sustain a fire depends on the difference between the ignition point and the temperature of the surroundings. Heat is lost to the surroundings by a combination of radiation and convection. The colder the surroundings, the greater the rate of heat loss via these processes. Therefore, for low-grade fuels, such as wood, that do not produce much heat when they combust, a fire would not be able to sustain itself if the surroundings were cold enough. Instead, to keep it burning you would need a continual supply of heat from an external source.

However, there is a limit to how cold you can make the surroundings, culminating at absolute zero. So fuels with a high enough heat of reaction will always be able to sustain a fire, no matter how cold the surroundings. Conversely, even fuels that are normally difficult to ignite can be made to burn if the surroundings are hot enough.

DECEMBER

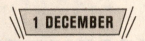

Does playing music in shops increase sales?

THESE DAYS SOME shops don't even wait for 1 December before piping 'All I Want For Christmas Is You' through the sound system. For those less festively inclined, it's enough to make you shout 'Bah, humbug!'

And yet music does increase sales – not only that, but specific music can increase sales of specific items. In a 1997 study, psychologist Adrian North and colleagues played stereotypically French and German music on alternate days in front of a display of French and German wines. French music led to French wines outselling German ones, whereas German music had the opposite effect. Yet a questionnaire suggested customers were unaware that the music had an effect on their product choices.

This is probably because the music creates a mental association with a product, although exactly why remains a mystery. It works with smells too – the piped smell of fresh, ground coffee or baking bread also increases sales.

The impact of background music is less clear, though many retailers feel that a library-like atmosphere of silence can be off-putting. So, despite the cost (in the UK, payments must be made to the Performing Rights Society), such music is widely played.

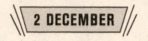

Why do tigers have stripes?

THE BEAUTIFUL STRIPED markings on tigers' coats are unique in the big cat family. Other closely related big cats have spotty rosette or cloud-shaped body markings (leopards, jaguars and clouded leopards), or plain coats (lions).

Work by a team at the University of Bristol has shown that cat patterning evolved to provide camouflage suited to the cats' particular habitats and behaviours, enabling them to capture prey more effectively (and escape predation in the case of the smaller cats). In general, plainer species such as lions live in open environments and hunt by day, whereas cats with complex patterning like leopards and tigers have more nocturnal habits and live in environments with more trees.

Unfortunately, with no other striped big cats around besides tigers, we cannot use the same methods to identify the evolutionary factor which drove tigers to depart from the ancestral big-cat pattern. Tigers are much bigger than jaguars and leopards, but in general they have a similar ecology, and tigers' historical range and habitat overlap considerably with those of leopards. So why don't they look similar?

One idea put forward decades ago, but for which evidence is still lacking, is that compared with the typical leopard habitat, the average tiger habitat contains a lot of vertical features such as bamboo. Quantifying this would be straightforward except that with the tigers' range now so shrunken, it is hard to know exactly what sort of forest their coat evolved in. Tigers are obviously well camouflaged, yet the factors behind their appearance remain an enigma.

Intriguingly, the Bristol team has also shown that big-cat patterning changes relatively rapidly over evolutionary timescales. One day our descendants might wonder at the beauty of striped leopards and spotty tigers.

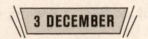

Why is ice slippery?

FOR PHYSICISTS NO less than figure skaters, ice is remarkably hard to get a grip on. The overwhelming consensus is that ice has low friction because of a thin film of liquid water coating its surface. Hence skaters balanced on thin metal blades can glide smoothly across the ice rink, but grind to a halt on the wooden floor beyond. The tricky part is how this liquid layer forms. More than a century of research has brought us little closer to a definitive answer.

In June 1850, Michael Faraday told an audience at London's Royal Institution that pressing two ice cubes together led to them forming a single block. He attributed this to the appearance of an intervening film of water that quickly refreezes. For many years, the appearance of this layer of water was put down to pressure. In fact, even a person of above-average weight on a single skate generates far too little pressure to account for the observed melting.

Today there are several theories. According to Changqing Sun of Nanyang Technical University in Singapore, the assumption that the slippery layer coating ice is a liquid is fundamentally flawed. He says this layer should properly be called a 'supersolid skin' because the weak bonds between H_2O molecules at the surface are stretched, but unlike in liquid water none of them are broken.

He also argues that this elongation of bonds ultimately produces a repulsive electrostatic force between the surface layer and anything it comes into contact with. He compares the effect to the electromagnetic force that levitates Maglev trains, or the air pressure a hovercraft generates beneath its hull. If he's right, his model helps to explain many of the layer's properties, including its remarkably low friction.

Most in the ice field are not convinced. Gen Sazaki at Hokkaido University in Sapporo, Japan, who made the first direct observations of this layer in 2013, prefers to call it a quasi-liquid. He thinks it represents a transitional stage between solid and liquid as the temperature goes up.

For Sazaki, understanding how this mysterious sheet of H_2O forms is still some way off. Even when it comes to something as familiar as slipping on ice, he says, 'reality is much more complicated than we expected'.

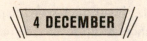

4 DECEMBER

Do doctors live longer?

DOCTORS ARE AS interested in their own mortality as any other group of people, and have investigated this very subject.

The authors of a 1997 article in the journal *Occupational and Environmental Medicine* looked at the cause of death listed in UK Department of Health records for more than 20,000 hospital consultants who died between 1962 and 1992. Overall, these doctors had rates of lung cancer, heart disease and diabetes that were less than half those for the general population. The study also found distinct statistical variations depending on the consultant's area of specialisation. For instance, suicide was more common among anaesthetists than other doctors.

According to this study, doctors do indeed live longer than the average for their corresponding national population, but whether this is because of their knowledge of medicine and proximity to medical services is another matter. After all, it is well known that socio-economic status and educational attainment have a big impact on health and hence life expectancy. A study of the UK population showed that middle-class professionals such as doctors and accountants outlive builders and cleaners by eight years on average.

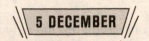

5 DECEMBER

Why do flying fish fly?

STRICTLY SPEAKING, THE flying fish does not fly, it indulges in a form of powered gliding, using its tail fins to propel it clear of the water. It sustains its leap with high-speed flapping of its oversized pectoral fins for distances of up to 100 metres.

The usual explanation of this activity seems to be to escape predators. If one can manage to tear one's eyes away from the magic of the unexpected and iridescent appearance of a flying fish, a somewhat more substantial fish can often be seen following its flight path just below the surface.

It has been suggested that their glides are energy-saving, but this is very unlikely as the vigorous takeoffs are produced by white, anaerobic muscle beating the tail at a rate of 50 to 70 beats per second, and this must be very expensive in terms of energy use.

Flying fish have corneas with flat facets, so they can see in both air and water. There is some evidence to suggest that they can choose landing sites. This might allow them to fly from food-poor to food-rich areas, but convincing evidence of this is lacking.

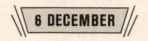

Why do balloons spiral when released?

FOR A BALLOON to fly in a straight line, the direction of the jet of expelled air would have to be in line with the balloon's centre of mass and its centre of drag – the point where the forces resisting the balloon's forward motion are symmetrical. If these two centres don't coincide, the centre of drag should be behind the centre of mass, otherwise stability is compromised.

If the balloon's line of thrust does not pass through the centre of mass (which is almost certain) but is in the same plane as the line joining the centres of mass and drag (which is unlikely), the balloon would travel in a circle in that plane, although the pull of gravity will ultimately force it down to the ground, especially as the air driving it forward expires.

However, because these lines generally do not intersect, thrust from the balloon's opening comes at an angle to the plane of the circle, pushing the balloon into the helical, screw-like motion. The thrust of the balloon and the air resistance to the balloon will not cancel each other out in such a situation and so a turning moment is exerted.

Balloons are not finely produced pieces of high-precision engineering, so it is likely that the line of thrust will never be near the centre of drag and the resulting torque will always make the balloon spiral wildly. However, you can aim the balloon to a certain extent by taping a nozzle to it at different angles (try cutting one from lengths of drinking straw). With trial and error, you may be able to get the balloon to fly in a reasonably straight line.

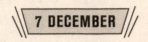

Why is wee always the same sort of colour?

NO MATTER WHAT colour of drink you consume, when the liquid finally leaves the body the colour has gone. Coloured substances in drinks (or food) are usually organic compounds that the human body has an amazing ability to metabolise, turning them into colourless carbon dioxide, water and urea. The toughest stuff is taken care of by the liver, which is a veritable living waste incinerator, while the kidneys take care of removing waste products from your blood. By the time it leaves your body the liquid is almost unrelated in chemical composition to the original liquid you consumed.

Any substance, solid or liquid, that goes down your oesophagus and passes through the digestive tract, if not absorbed, is incorporated into faecal matter. Urine, by contrast, is produced by the kidneys from metabolic waste produced in the tissues that has been transported through the bloodstream. This waste is added to any excess water you have consumed to produce urine of various shades of yellow. This is stored in and passed out from the bladder.

Any coloured compound that you drink either will or will not interact biochemically with the body's systems. If it does, this interaction (like any other chemical reaction it might undergo) will tend to alter or eliminate its colour. If it does not, the digestive system will usually decline to absorb it and it will be excreted in the faeces which shows considerably more variation in colour than urine.

Some coloured substances can make it through your digestive system and into your urine. This occurs when the intake of coloured substances exceeds what the body can quickly metabolise and the colouring is not removed as the liquid leaves the body. Anybody who wants to see this effect should consume a large quantity of borscht (beetroot soup). You'll notice that your urine takes on a distinctly pinkish hue.

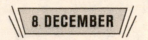

Why do millipedes have so many legs?

MILLIPEDES AND EARTHWORMS have similar lifestyles. Both burrow in soil, eating dead and decaying vegetation, but they have evolved very different methods for forcing their way through the soil. Worms use the strong muscles in their body walls to build up pressure in the body cavity, and so develop the forces needed to push forward or widen a crevice in the soil. Millipedes, however, use their legs to push through the soil. The more legs the animal has, the harder it can push.

Millipedes are different from centipedes. They have very large numbers of short legs because long legs would be a liability in a burrow. Centipedes, which spend their time on the surface or among leaf litter, have fewer, longer legs. They have little need to push, but have to run faster than millipedes.

Millipedes, centipedes and earthworms all have long, slender bodies, divided into large numbers of segments. Except at the two ends of the body, all the segments are built to more or less the same design. Similarly, many products of human engineering are built largely from a series of identical modules. For example, identical seats and windows are repeated many times along the length of a bus. The advantage of this is that one design will serve for all the seats, and one machine can make them all.

In the same way, the repetition of segments in animals reduces the quantity of genetic information needed for development. Millipedes presumably evolved from an ancestor with fewer segments and correspondingly fewer legs, simply by changes in the genes that specify the number of segments.

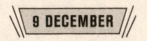

9 DECEMBER

Why do we have fingerprints?

FINGERPRINTS AND THEIR unique patterns are a feature of any police procedural TV series, but that's not exactly an evolutionary benefit. So what purpose did they evolve to serve?

Fingerprints are the visible parts of rete ridges, where the epidermis of the skin dips down into the dermis, forming an interlocking structure (similar to interlaced fingers). The unique patterns are simply due to the semi-random way in which the ridges and the structures in the dermis grow.

Fingerprints help us in gripping and handling objects in a variety of conditions. They work on the same principle as the tyres of a car. While smooth surfaces are fine for gripping in a dry environment, they are useless in a wet one. So we have evolved a system of troughs and ridges, to help channel the water away from the fingertips, leaving a dry surface which allows a better grip. Fingerprints protect against shearing (sideways) stress, which would otherwise separate the two layers of skin and allow fluid to accumulate in the space (a blister). They appear on skin surfaces which are subject to constant shearing stress, such as fingers, palms, toes and heels.

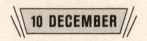

How do car airbags work?

AIRBAG COVERS ARE formed from moulded plastic and have lines built into them that are much thinner than the rest of the cover. When the airbag inflates, it forces its way through the covers, which fracture along these very thin lines. Obviously, it is important that the cover does not become a projectile, so it also has other thin sections which act as hinges. These hinges ensure that the fractured sections of the cover rotate harmlessly away from the occupant, rather like a pair of barn doors swinging open. These thin hinge lines and fracture lines are often visible, sometimes looking like a large 'H' (especially in older or less expensive cars).

Airbag engineers pay particular attention to the design and fixing of any logos or manufacturers' badges which are fitted to the cover in the centre of the steering wheel to ensure that the badge remains attached to one of the doors formed by the opening of the cover, instead of coming loose and causing injury.

In the case of side-impact airbags contained within the seat, a similar effect is achieved by providing a weakened seam of stitching in the seat cover, immediately alongside the airbag. This is one good reason not to fit seat covers.

Airbag engineers also have to ensure that the rapidly unfolding envelope of the bag moves straight towards the person it is supposed to protect, rather than across the driver's or passenger's face and chest. This involves very careful analysis of folding patterns, predictive software and analysis of high-speed photography from test firings. Engineers have also learned a great deal from origami.

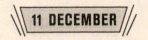
Does the length of a day change?

THE MOON ORBITS Earth once a month, and Earth spins on its axis once a day. The effect of the tides is like a brake, causing the spin of Earth to slow while, to conserve angular momentum, accelerating the moon to a higher orbit.

As a result, the length of a day is slowly increasing by approximately 2 milliseconds every century. This has been measured by examining eclipse records dating back to 600 bc, because the position at which an eclipse can be observed is a sensitive indicator of how far Earth has turned between that time and today. Although this lengthening of a day is imperceptibly small for most practical purposes, the mismatch between the length of a day now and a day 100 years ago is sufficiently large to bring the time recorded by atomic clocks out of step with astronomical time. Every few years it is necessary to introduce leap seconds to synchronise them.

Interestingly, in planetary systems where the moon orbits in the opposite direction to the planetary spin, the effect of tides is to slow the moon down and its orbit lowers, presumably until it is dragged into the planet. At least Earth's future generations don't have that to worry about. Laser ranging to the moon has also shown that the moon is moving away at about 4 centimetres per year.

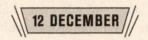

How many organisms live on the human body?

THE MICROORGANISMS THAT inhabit the body of a healthy human being are known as the normal microbial fauna and they come in two different types – those that are permanently resident and those that are transient. Of course, any number of fascinating and nasty parasites can join this microbial community and make the human body their home.

In his seminal work *Life on Man*, bacteriologist Theodor Rosebury counted 80 distinguishable species living in the mouth alone and estimated that the total number of bacteria excreted each day by an adult ranges from 100 billion to 100 trillion.

Microbes inhabit every surface of a healthy adult human that is exposed to the outside, such as the skin, or that is accessible from the outside – the intestines, from mouth to anus, plus eyes, ears and airways. Rosebury estimates that 10 million individual bacteria live on the average square centimetre of human skin, describing the surface of the body as akin to a 'teeming human population during Christmas shopping'.

However, this figure can vary widely throughout the almost 2 square metres that make up the surface area of a human. In the oily skin that is found on the side of the nose or in a sweaty armpit, the figure can increase tenfold, while on the surface of the teeth, throat or alimentary tract, these concentrations can increase a thousandfold. These inside surfaces are the most densely populated region of the human body. Conversely, on those surfaces where there is liquid flow removing bacteria, such as the tear duct or genito-urinary surfaces, the populations of organisms are much thinner. Indeed, Rosebury could detect no microbial life at all in the bladder and lower reaches of the lungs.

Yet, while the figures appear huge, he estimates that all the bacteria living on the external surface of a human would fit into a medium-sized pea, while all those on the inside would fill a vessel with a capacity of a mere 300 millilitres.

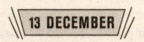

When is a submarine not a submarine?

THE ANSWER, AS an embarrassed Swedish navy had to admit in 1996, is when it's a mink or an otter.

What the navy thought was the sinister sound of Soviet propellers was, in fact, the furious paddling of little legs. A scientific commission set up by the government and chaired by the former director of the Swedish Engineering Science Academy, Hans Forsberg, concluded that most of the invading submarines reported by the navy were mythical. Of more than 6,000 reports of 'alien underwater activity' between 1981 and 1994, the commission found firm evidence for only six incidents.

In every other case the evidence, often based on sightings by the public, was unreliable.

On 40 occasions between 1992 and 1994, a defence network of microphones attached to buoys detected the sound of bubbles caused by a rotational movement in the water. The navy estimated the speed at up to 200 revolutions per minute, and assumed it must be submarine propellers. But the navy was wrong. According to the commission's secretary, Ingvar Akesson, tests with swimming mink or otters showed that they could produce the same readings as propellers.

Why does honey make bread concave?

TAKE A FRESH, untoasted slice of bread and spread it with honey. Then set it aside. If you leave it for a few minutes, you'll see that it becomes concave on the side spread with honey.

For some of you, your bread might not have time to go concave at breakfast. However, for those folk who chomp their honeyed bread in a more leisurely fashion, there is a simple explanation.

The bread becomes concave because bread is approximately 40 per cent water, while honey is a strong solution containing approximately 80 per cent sugar. Sugar is hygroscopic, which means that it soaks up moisture. This causes moisture to be drawn out of the bread and into the honey by osmosis. Extracting the water makes the bread shrink, but only on the side exposed to the honey. This explains why the bread becomes concave.

This is less likely to happen if you butter your bread before spreading the honey. Butter forms a fat-rich, water-impermeable layer that protects the bread from dehydration by the honey.

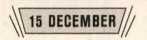

15 DECEMBER

Do dogs know they're dogs?

THIS QUESTION CALLS to mind cyberneticist Stafford Beer, writing in a 1970s edition of *New Scientist*: 'Man: "Hello, my boy. And what is your dog's name?" Boy: "I don't know. But we call him Rover."'

The boy's reply reveals his belief that his dog has a mental image of itself (which he assumes to include a name), but at the same time confesses his inability to penetrate the dog's psyche. And he's right: the short but unsatisfying answer to the question above is that we don't know what goes on in a dog's mind.

We can, though, make a reasonable stab at it. Canids in general start to develop social relationships when their eyes and ears open at about two weeks of age. During the critical period between 2 and 16 weeks, puppies learn the social rules that will shape their behaviour for the rest of their lives, including recognition of conspecifics and appropriate mates.

After 16 weeks, this period of rapid learning and adaptation ends, and the social skills the dog has are pretty much set for life. This is why it is so important for puppies to have intimate contact with people from the time they are born. Traditionally, we adopt pet dogs when they are eight or nine weeks old, right in the middle of this period of social development, and proceed to lavish them with attention and experiences through to the end of that 16-week period. The result is that dogs see nothing at all odd about their tall, hairless pack-mates.

Likewise, livestock-guarding dogs, such as those protecting sheep, are trained for their jobs by removing them from their mothers at just a few weeks of age and allowing them to grow up with sheep as their companions. The sheep are then forever recognised as family and are socialised with and protected as such.

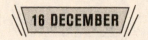

Will Mount Everest always be the tallest mountain on the planet?

THIS IS A good question, but the short answer is that nobody knows. Let's clarify some terms first: Mount Everest is not the tallest mountain on the planet by a long way – that accolade goes to Mauna Kea in Hawaii, although much of it is below the ocean. The summit of Everest is not even the furthest point from the centre of the planet – that is believed to be the summit of Chimborazo in Ecuador. The summit of Everest is, though, the highest point on the planet above mean sea level.

It seems to be generally accepted that the Himalayas, in which Everest is located, are still rising by about 5 millimetres per year as a result of the movement of tectonic plates in the region. Geological time is measured in millions of years, so the summit of Everest is likely to remain the highest point on the planet for many years yet.

Other relatively new ranges are rising too, but, because the Himalayas include most of the top 100 highest points on the planet, it's unlikely that Everest – more than 200 metres taller than its nearest rival – will be overtaken for some time.

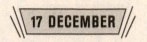

Why are there so many ingredients in shampoo?

READ A LABEL on a shampoo bottle and the list of ingredients is mind-boggling. But believe it or not, most of the ingredients in shampoo are doing nothing to your hair. Only the detergents clean it, while the rest of the ingredients are used to improve the appearance, smell, texture and shelf life of the product. Providing there are no restrictions on the use of an ingredient, manufacturers are free to use just about anything they like in a cosmetic or toiletry.

A typical shampoo is mostly water, containing between 5 and 20 per cent detergent, with shampoo for dry hair containing less detergent than shampoo for greasy hair. The most widely used detergent is sodium lauryl sulphate (SLS), but as this gives a poor lather, sodium laureth sulphate, which produces a stronger foam, or foam boosters such as cocamide DEA or cocamidopropyl betaine, may also be present. Lather plays no part in the detergency process, although it does keep the detergent and any other active ingredients close to your hair and scalp. Thick lather also reinforces the psychological link between the shampoo and perceived cleaning power.

Oils can be added to counteract the drying effect that detergents have on hair and, therefore, emulsifiers and emulsion stabilisers must also be added. The oils can be anything from natural vegetable oils to synthetic silicone polymers such as methicone and dimethicone. These also have a conditioning effect, helping to smooth the cuticle layer of the hair shafts.

At least two preservatives are normally present. One must be water-soluble to protect the watery part of the shampoo, and the other oil-soluble to preserve the oils in the emulsion. The paraben family, formaldehyde, glyoxal and the methylchloroisothiazolinone/

methylisothiazolinone mixture are all preservatives commonly added to shampoos.

Thickeners are added to adjust the texture and pouring properties of the product, colourants impart the required colour, UV-absorbers stop the colours from fading, opacifiers give the shampoo a creamy or pearlescent appearance, and natural or artificial fragrances make it smell nice.

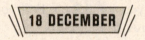

18 DECEMBER

How long is now?

THE FEELING OF existing permanently in a present moment as time flows past us is one of the most essential of human experiences. The 'now' it defines is a point in time that has no duration, just a personal portal between past and future. But that sort of now is too abstract even for physicists. Time does not flow – it is a dimension like space that just *is*, and all of it exists for all time.

This 'block universe' thinking is made necessary not least by Einstein's relativity, which says that now is, well, relative. Take a photon from the cosmic microwave background. It reveals to us a picture from just after the big bang 13.8 billion years ago. But from the photon's perspective, its journey from back then takes no time at all: our now, the now of the big bang and all the nows in between happen simultaneously. From a cosmic perspective you can define now to be as long as you like – your answer will be equally right (or wrong).

But if the present is just an individual illusion created in our brains, we can still ask how long *that* now is. Marc Wittmann of the Institute for Frontier Areas of Psychology and Mental Health in Freiburg, Germany, thinks there are broadly three answers.

First comes the 'functional moment'. This is the time we need between two stimuli – a click in the right and left ears, say – to distinguish their order. It varies between senses, but is generally a few tens of milliseconds.

Our brains need a little longer to stitch various stimuli together into a conscious 'experienced moment'. Various experiments show that this lasts between two and three seconds. Listen to a metronome, for example, and you will have no trouble tapping your finger or foot to its beat provided the next comes within this time frame. Any longer, however, and you are likely to hear yourself counting under your breath to synchronise your beats.

Our individual experienced moments are then further stitched together by our working memories to produce a third now – the sense of continuity and time passing. The length of this now is tens of seconds, although it varies considerably depending on the density of stimuli our brains are working with. 'The feeling that time drags – you're waiting for the bus, you're cold, your smartphone isn't working – that's very real,' says Wittmann. Conversely, our brains ratchet up the processing speed if we're exploring new places or having new experiences.

This is the central truth of now, as far as there is any. 'Time and self are intricately linked,' says Wittmann. So perhaps the only valid answer is to live in the now – however long that is.

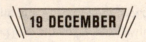

Why do butterflies flutter?

APART FROM FLIGHT, butterflies use their wings for temperature control, adjusting themselves to shed or absorb warmth. These actions tend to be conspicuous, which creates selective pressures for species to advertise themselves as vividly as possible to their own species, or warn predators that they are unpleasant to eat.

Such selection favoured flat, exaggeratedly conspicuous wings, which in turn constrains styles of flight, dictating larger, slower flapping movements than the compact, high-frequency wing vibrations of their ancestral moths.

The visual impression that butterflies' style of flight conveys is important; they emphasise it by up-and-down flapping of their billboards, and by individual styles of aerial dancing. This advertises their quality as mates, and provides recognition clues to their kind. For example, long-distance migratory butterflies such as monarchs don't fly like gliders in jungle clearings, or like clouded yellows dancing and dipping over alfalfa fields to attract mates, or leisurely *Acraea* species that only stupid birds taste twice.

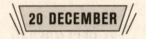

20 DECEMBER

How do snowflakes form?

THE BEAUTIFUL SIX-POINTED designs of snowflakes lend them a magical quality – but how do they form their distinctive shapes?

The truth is there is no spooky power at work here. Ice forms into hexagonal crystals because of the six-fold symmetry of the crystal lattice it forms when it is at or below freezing at a pressure of 1 atmosphere (a number of other ices form at extreme pressures that dance to different lattice tunes).

The crystal itself is concerned only with finding a molecule that might fit a gap in its lattice, which is how crystals grow. If the molecule is moving too fast, it will not settle into the gap. It is much like the ball on the roulette wheel that bounces here and there until it has lost enough energy to be unable to escape from one of the numbered indentations. Similarly, the water molecules bounce and jostle until one has lost enough energy to settle into the gap in the lattice.

In an environment where heat is slowly being withdrawn, molecules will fill the lattice gaps in an orderly manner and the snow crystals we know and love will grow with that six-fold symmetry.

Differences in temperature and pressure in the micro-environment of each snowflake will create enough variation to ensure that the crystals will not actually be identical, but they can be very similar.

A high-speed freeze will result in a dense mass of interlocked crystals that create ice with no apparent crystalline structure. This is seen in fast-frozen ice cubes, which appear clear. But in a slow-freezing, 'constrained' system, such as on a pane of glass or a pond surface, gradual cooling will create the typical hexagonal crystal pattern at the advancing edge of crystallisation.

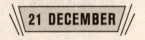

How fast would you need to travel to feel heat from air friction?

OBJECTS THAT ENTER Earth's atmosphere are subject to extreme heat through air friction – so much so that returning spacecraft have to be very well insulated, and rocks and small asteroids burn up.

To feel the heat of air friction, cyclists would have to break a lot of records and a lot of bikes. Even at about 900 kilometres per hour, the skin of a cruising commercial jetliner only becomes about 30 °C warmer than the air in the upper troposphere – still well below freezing.

At supersonic speeds, aircraft do get much hotter. Concorde routinely got too hot to touch: more than 120 °C near the front and more than 80 °C towards the back. In fact, its maximum cruising speed was limited not so much by its power as by the temperatures that its aluminium skin could tolerate. Such temperatures distorted its airframe, but not as badly as the Mach 3 Lockheed Blackbird, which reached well over 300 °C, requiring special engineering and construction materials to cope with the expansion the plane underwent as it heated up during flight.

Because the Blackbird lacked sealants that could withstand operational temperatures, the plane actually leaked fuel until it warmed up after take-off. Fortunately, as a cyclist you will not feel frictional warming below speeds at which the slip-stream would take your skin off.

22 DECEMBER

Why do hens' eggs vary in colour?

SHORTLY BEFORE THEY are laid, hens' eggs are white. The brown pigmentation associated with breeds such as the Rhode Island red and the maran is a last-minute addition during egg formation and, like a fresh coat of paint, can come off surprisingly easily.

More than 90 per cent of the shell of a hen's egg consists of calcium carbonate crystals bound in a protein matrix. The shell starts to form after the egg has reached the uterus, where it stays for around 20 hours prior to being laid.

During this time, glands secrete the shell around the membranes that hold the yolk and albumen. In brown-egg-laying breeds, the cells lining the shell glands release pigment during the last 3 to 4 hours of shell formation. Most of the pigment is transferred to the cuticle, a waterproof membrane that surrounds the porous eggshell.

Several factors can disturb the cuticle formation process and thus pigmentation, such as ageing, viral infections – including that perennial chicken farmer's nemesis, bronchitis – and drugs such as nicarbazin, which has been widely fed to poultry to combat a disease caused by a type of protozoan. Possibly the most significant factor affecting egg pigmentation is exposure to stress during the formation of the egg.

If a flock of hens is disturbed by a fox during the night, for example, they might well lay paler eggs in the morning. The adrenaline the hens release puts egg-laying on hold and shuts down shell formation. The egg's pigmentation will be affected if the cuticle doesn't form properly.

The issue may have some significance for public health. The waterproof cuticle is the egg's defence against bacteria. As shell colour is affected by how well the cuticle forms, it also provides a visual test of how free from harmful bacteria an egg may be.

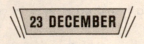 **23 DECEMBER**

What is the Christmas tree smell?

THAT CHRISTMAS TREE smell is the scent of coniferous evolution. Over millions of years, these trees have equipped themselves with a cocktail of chemical weaponry, including substances that act as fungicides and bactericides that deter herbivorous pests, large and small.

The chemical combination varies with tree species, but generally consists of a mix of aromatics, including terpenes such as alpha- and beta-pinene, limonene and camphene; and esters such as bornyl acetate. By lucky happenstance, we tend to find these scents appealing – so much so that they are added to perfumes and air fresheners.

If you happen to have an artificial tree, you will smell a different kind of ester, probably a phthalate or suchlike, used to make the plastic fronds softer and more flexible. If so, soak a few tree decorations in pine-scented disinfectant to give a conifer fragrance to your holiday season.

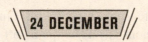

What time is it at the North Pole?

THERE ARE TWO answers to this question. The first is that the time for a person is the time determined by her or his circadian rhythm. Initially, this physiological time will be close to the time for the longitude where the individual lived before visiting the pole. Over a period of weeks at the pole, this time will drift as the individual settles into a rhythm with a period that is usually about 25 hours long.

Time, from a geophysical point of view, however, is related to the position of the sun over the Earth and to the position of the observer. Because any direction from the North Pole is south, the sun is always in the south and whatever the time is at the North Pole, it is always the same time.

What time is that? The International Date Line runs through the North Pole, leaving the pole sitting eternally between one date and the next. In other words, it is always midnight at the North Pole.

This, of course, explains how Father Christmas manages to deliver presents to every good little boy and girl throughout the world in the space of a single night.

He just heads out of his grotto due south (which from the North Pole is any direction), drops off as many presents as he can fit on his sleigh and then he heads back home where it is exactly the same time as when he left. So he can then drop off more prezzies, return home, and so forth.

**25 DECEMBER**

Why do geese have so much fat?

IF YOU'VE CHOSEN to cook a goose for Christmas, you may be astounded at the amount of fat that pours off it. Why do they need so much?

In the wild, geese are aquatic birds, with many species being migratory. They consequently need both a substantial energy reserve to sustain them during the long flights of their migration, and good insulation to protect them from the cold and wet. Fat covers both necessities quite nicely.

Goose fat, therefore, is by no means simply dead weight. An adult greylag (the species from which almost all breeds of domesticated goose descend) can weigh up to 5.5 kilograms. With a wingspan of over 160 centimetres, that makes for a relatively low wing loading – the ratio of mass to wing area.

In aviation terms, geese could be considered the 'long-haul wide-bodies' of the bird world, and a fuel load that can sustain them over thousands of kilometres is vital rather than a burden.

Of course, the ratio of goose fat to body weight is much lower in a wild bird than in the domesticated fowl typically eaten for Christmas dinner. For a number of reasons, domestic geese have had their body fat augmented by a combination of selective breeding and diet.

Prior to the introduction of the railways, from a farmer's perspective the goose had quite a big advantage over the turkey – it was a lot easier to transport. Goose farmers based in East Anglia in the UK, for example, would simply walk their geese to London's Smithfield market. They knew that, unlike turkeys, a goose could not roost overnight up a tree from which it would be almost impossible to retrieve the following morning. Furthermore, geese could sustain themselves on this 160-kilometre waddle by grazing on grass, drawing on their reserves of fat for additional sustenance.

From a cook's perspective, the goose's advantage over the turkey was that it required no additional basting, its own supply of body fat being sufficient if spooned over the bird periodically while roasting. Indeed, so much fat comes off a full-grown bird that the excess can be used to baste other meat cooked at the same time, a favourite culinary trick of Charles Dickens.

Through his writing, Dickens helped make turkey an essential part of the British Christmas dinner. Ironically, he preferred the taste of goose, and to further his enjoyment of the bird as part of his seasonal lunch he would roast a whole ox heart in a pan placed underneath the trivet on which the goose was cooking, so that the heart's bland but succulent meat would soak up the goose's flavour, along with its fat as it dripped from above.

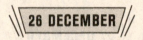

26 DECEMBER

How do fruitcakes last so long?

AFTER BAKING A fruit cake for Thanksgiving in the late 1870s, Fidelia Bates of Tecumseh, Michigan, promptly died. This presented a rather delicate question for the family: who would be the first to eat a piece of the dead woman's cake?

As it turned out, Mrs Bates's family has resisted temptation to polish it off for over 140 years, and counting. The cake has been kept under a glass lid and stored high up in a cupboard ever since, save for the occasional appearance on TV or at Morgan's grandchildren's school show-and-tell.

If a daredevil were to try the cake, they would find it still edible thanks to the sugar content. Sugars are hygroscopic, that is, they draw water from their surroundings, including from any bacteria or fungi. This prevents growth of such microbial contaminants. Bees process honey for exactly the same purpose, reducing the water content of the collected nectar to around 20 per cent so that the honey, which is 80 per cent sugar, can resist microbial spoilage.

Jam – a method of preserving fruit in sugar – is about 60 per cent sugar. Traditional fruit cake can be around 60 per cent carbohydrate, with the dried fruit and sugar together contributing about three-quarters of that. In other words, fruit cake can be as much as 50 per cent sugar.

In the days before refrigeration this was an important means of preparing food for storage or travel, as well as providing a sweet delicacy. In Europe prior to the introduction of sugar from sugar cane in around 1100, dried fruits and honey were the only such preservatives, hence the ubiquity of fruit cake, fruit mince and the like in post-medieval European cookery for special occasions.

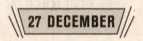

27 DECEMBER

Why is it colder at the South Pole than at the North Pole?

MUCH OF THE temperature difference between the two poles can be explained by their difference in elevation. The North Pole (with monthly average temperatures in winter of around –30 °C) lies on sea ice on the surface of the Arctic Ocean while the South Pole (at around –60 °C) is 2800 metres above sea level on the ice sheets of the Antarctic continent.

The background variation of temperature with height (in Antarctica about –6 °C per kilometre gain in height) thus accounts for over half the difference. Also, the 'thinner' (and hence colder, drier and less cloudy) atmosphere overlying the South Pole reflects less heat back to the surface than its northern counterpart. Much of the remainder of the temperature difference can be explained by the contrasting atmospheric circulation regimes in the two hemispheres.

The continents of the northern hemisphere drive quasistationary 'planetary waves' in the atmosphere. These waves transport heat polewards and also 'steer' mid-latitude depressions into the north polar regions. The continents of the southern hemisphere are smaller and lower than those in the north, so the southern hemisphere planetary waves (and associated heat transport) are smaller.

The high mountains of Antarctica also block the poleward movement of mid-latitude depressions, which rarely penetrate into the interior of the continent. Finally, the atmosphere at the North Pole receives some heat from the underlying Arctic Ocean. Although the heat conducted through the 2 to 3 metres of sea ice that typically cover the ocean is small, large amounts of heat can be exchanged over the narrow 'leads' of open water that occasionally form between ice floes.

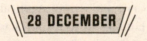

28 DECEMBER

Does drinking a lot of water detox the body?

THERE IS A current fad for drinking water in order to purify ('detox') the system after an excess of anything from alcohol to red meat to Christmas dinner.

But evidence suggests this is unnecessary. The minimum volume of urine required by the kidneys to excrete the waste products of the human metabolism is about half a litre a day. Because we lose another half litre through breathing, sweating and defecating we need to replace the total lost by drinking about 1 litre a day.

Drinking more than this merely dilutes the urine – the same amount of waste products (or 'toxins' as they are popularly known) are still removed from our bodies. Coffee and tea have a mild diuretic effect, but your body still gets a net fluid gain from drinking them – you do not urinate more liquid than you consume. Alcohol, on the other hand, is a much stronger diuretic, so you should always drink lots of soft drinks or water after over-indulging.

'Detoxification' may be popular, but there is no evidence to support the argument that our kidneys cannot cope with the input and output of normal or even excessive consumption as long as the latter is not on a long-term basis. Drinking more water than we need makes no significant difference to the elimination of waste products from our bodies.

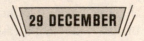

29 DECEMBER

Why does beer foam when poured?

POUR SOME WINE or beer rapidly into a vertical glass so that it froths up to the rim, let the bubbles subside and then add more. You'll see that the second batch of liquid will not have the same frothing effect as the first. Why?

Beer, sparkling wine and other fizzy drinks are supersaturated with gas. Although the laws of thermodynamics favour the gas bubbling out of the dissolved state, bubble formation is unlikely since bubbles must start small. Because the pressure of these tiny bubbles can reach about 30 atmospheres in a bubble only 0.1 micrometres in diameter and the solubility of gases increases with increasing pressure (Henry's law), the gas is forced back into solution as quickly as it comes out.

However, bubbles can form around dust particles, surface

irregularities and scratches. These areas, know as nucleation sites, are hydrophobic (they repel water) and allow gas pockets to form without first forming the tiny bubbles. Once the gas pocket reaches a critical size, it bulges out and rounds up into a properly convex bubble, the radius of the curvature of which is sufficiently large to prevent the self-collapse described earlier.

Then there is a cascade effect. If the bubbles reach a certain critical number per unit volume, this in itself constitutes a physical disturbance and results in the release of yet more bubbles.

If you want to prevent foaming, wet the inside of the glass. Once the inside of the glass is wet, the imperfections will no longer function as centres of nucleation. Most of the dust particles and all of the scratches will, of course, still be there. However, these will have been coated with liquid and the fresh carbonated liquid will reach them very slowly, by diffusion. Bubbles will still be produced, but at a rate that is too slow for the cascade effect to come into play. As a result, the drink will not froth over.

Thanks to modern production techniques, today's glasses are of such good quality that some manufacturers build in deliberate imperfections, especially in beer glasses, in order to generate enough bubbles to maintain the head on the top of your tipple.

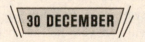

Why does snow under trees melt faster than snow in the open?

SNOW, LIKE EVERYTHING else, including apple trees, emits and absorbs radiation. While ultraviolet and visible radiation are strongly reflected (not absorbed) by snow, it is however a strong absorber of infrared radiation. The battle between the absorption and emission of radiation determines whether there is net warming or cooling of the snow – or neither.

So why would snow under a tree melt faster? At night, snow in the open absorbs infrared radiation from the ground and from the sky – which can be below -30 °C when it is clear.

Snow underneath a tree absorbs radiation emitted by the ground and by the tree, which is likely to be significantly warmer than the sky, so it emits more infrared energy.

This difference is sufficient to explain why snow underneath a tree might melt faster than snow of the same depth that is out in the open, and also explains why frost often does not form around trees. (It is also possible that shelter provided by trees when snow is falling leads to a thinner layer of snow there than in the rest of the immediate vicinity!)

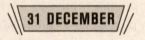

Which drinks cause the least hangover?

IT'S NEW YEAR'S EVE, and you might be considering the best way to avoid starting the new year with your best foot forward. If you're drinking, the best answer is not to drink – but in case you need a plan B, what drinks will give you the least bad hangover?

Most people consume alcoholic drinks for their ethanol content. However, many such drinks also contain amounts of other biologically active compounds known as congeners. Congeners include complex organic molecules such as polyphenols, other alcohols such as methanol, and histamine. They are produced along with ethanol during fermentation or the drink's ageing process.

Congeners are believed to contribute to the intoxicating effects of a drink, and the subsequent hangover. People who drink pure ethanol-based alcohols such as vodka have been shown to suffer fewer hangover symptoms than those who drink darker beverages such as whisky, brandy and red wine, all of which have a much higher congener content.

The congener denounced as the main hangover culprit is methanol. Humans metabolise methanol in a similar way to ethanol, but the end-product is different. Ethanol generates acetaldehyde, but when methanol is broken down, a major product is formaldehyde, which is more toxic than acetaldehyde and can cause blindness or death in high concentrations. Ethanol inhibits the metabolism of methanol, which may be why drinking 'the hair of the dog' can alleviate hangover symptoms.

Studies have found that the severity of different drinks' hangover symptoms decline in this order: brandy, red wine, rum, whisky, white wine, gin, vodka and pure ethanol.